Agriculture in India - Contemporary Challenges:
in the Context of Doubling Farmers' Income

Agriculture in India - Contemporary Challenges:
in the Context of Doubling Farmers' Income

Mohan Kanda

BS Publications
A unit of **BSP Books Pvt. Ltd.**
4-4-309/316, Giriraj Lane, Sultan Bazar,
Hyderabad - 500 095 - T.S.

Agriculture in India - Contemporary Challenges: *in the Context of Doubling Farmers' Income*
By Mohan Kanda

Published by:

MCR HRD Institute of Telangana,
Road No. 25, Jubilee Hills, Hyderabad 500033
Telangana (India)
www.cgg.gov.in
Phone: +91 40 2354 1907 / 09
Fax: +91 40 2354 1953
Email: info@cgg.gov.in

BSP **BS Publications**
A unit of **BSP Books Pvt. Ltd.**
4-4-309/316, Giriraj Lane, Sultan Bazar,
Hyderabad - 500 095
Phone : 040 - 23445600, 23445688
e-mail : info@bspbooks.net
www.bspbooks.net

ISBN: 978-93-91910-56-3 (Hardback)

प्रो. रमेश चन्द
सदस्य
Prof. Ramesh Chand
MEMBER
Tele.   : 011-23096774
Telefax : 011-23096756
E-mail : rc.niti@gov.in

भारत सरकार
नीति आयोग, संसद मार्ग
नई दिल्ली-110 001
Government of India
NATIONAL INSTITUTION FOR TRANSFORMING INDIA
NITI Aayog, Parliament Street
New Delhi-110 001

Foreword

Agriculture sector continues to dominate Indian economy with 17.7 per cent share in Gross Value Added (2019-20) and 44 per cent share in total workforce in the Country. The sector also assumes importance to attain SDGs 2030 not only to address hunger and nutrition greenhouse gas emissions and environment quality.

During the last three decades agri-food production has grown at a trend growth rate of 3 per cent per year while human population in the same period has grown at 1.6 per cent. Further, agri-food production maintains almost same growth rate but human population has shown deceleration in growth rate since 1981. The recent growth rate in population is estimated to be 1.4 per cent which is poised to fall further. This implies that growth in food production in India is likely to be almost three times the growth rate in population. Even now productivity level of most of crops and livestock is lower than world average and the country has potential to further raise trend growth rate by raising productivity. This sustained growth in agri-food production has helped Indian economy and the country in several ways and at the same time brought several challenges which are getting severe over time.

India, is now surplus in many commodities and need foreign market to sell increasing share of domestic production outside domestic market to prevent sharp fall in prices received by farmers. This requires improved efficiency in production and logistics. The growth rate in agriculture sector has been largely driven by output price support and input subsidies. They have caused serious distortions in output market and resulted in unsustainable use of natural resources. To some extent the country has been sacrificing long run for short run gains. This is evident from over exploitation of water resources and declining groundwater level in most parts of the country. This is linked to policies of free or highly subsidised power for irrigation and policy bias in favour of water intensive crops and irrigated agriculture. There are reports of farmers not getting remunerative prices as markets are not very competitive. Land lease laws in the country are such that they neither allow expansion of operational holding nor encourage exit from farming. The net result has been that agriculture suffers from poor competitiveness, low production efficiency, unsustainable use of natural resources, and low return to farmers.

The main reason for current state of agriculture is absence of required policy reforms and missing development initiatives in the sector for a long time. The reform agenda of 1991 left agriculture from its purview. The Central government has attempted three major reforms in agriculture to liberalize marketing of farm produce, facilitate partnership between private sector and farmers through innovative contract mechanism and changes in Essential Commodities Act to attract investments in storage, export and logistics. The book covers all above issues.

The author of this book provides a rich historical account of major agricultural events since the Bengal famine of 1770. The book discuss problems and developments of Agriculture sector in more details and present all interesting episodes since Independence. It presents a very able description of the need for change followed by the prescription to take agriculture forward.

There is hardly any aspect of agriculture development which is not covered in the book. As is required in this kind of work the author draws useful material available in writings of well known experts, government report and sources and present a well knotted documents. Beginning from Inputs used in agriculture the book goes on to discuss R&D in agriculture, extension, price policy, marketing, government intervention in agriculture, sustainability issues, climate change, risk in agriculture, logistics, producers group and contract farming. The author articulates the benefit of modern technology like precision farming, biotechnology, and GM technology for Indian farmers. As mentioned in its title the book emphasises policy and strategy for raising income of the farmers.

Like other experts on Indian agriculture the author points to the urgency of reforms in agriculture. The book is update and covers the current challenge of adjustment to shock of COVID 19 lockdown.

The coverage of information and depth of ideas presented in the book adequately reflects the long association and familiarity of Mohan Kanda with development and changes in Indian agriculture. The author is at his best in providing the preface of the book. It not only gives systematic chronological order of events around agriculture but also discuss significant changes in other fields of Indian economy.

The rich discussion is followed by equally impressive prescription in Chapter 4 which outlines way forward routed in the challenges and context described in the book. The book is written in a simple and persuasive style. I think it is worth reading and keeping in the shelves of students of agriculture, development planners and administrators. Common readers having interest in agriculture will find it handy to understand story of Indian agriculture - past, present and going forward.

Ramesh Chand
New Delhi.

Preface

"Everything else can wait", said Shri Jawahar Lal Nehru, India's first Prime Minister, "but not agriculture". More than seven decades have passed since those words were spoken. The fact of the matter is that agriculture is still waiting for its "tryst with destiny" – to use Nehru's own words again. Seven decades since it became a Republic, India's accomplishments make an impressive list. Achieving independence from colonial rule through a completely peaceful and non-violent movement was, in itself, a significant event in the history of the modern world. Achieving food security and nutritional adequacy (at least in the global sense), through a veritable rainbow of revolutions such as the Green Revolution in food grains, the White Revolution in milk production, the Yellow Revolution in oilseeds production, and the Blue Revolution in fish production was another spectacular achievement, helping, as it did, the country's transition from the days of the so called "ship to mouth" existence to an era of almost embarrassing surpluses. The stellar contributions made by towering personalities such as Dr. M.S. Swaminathan, Dr. Norman Borlaug and Dr. V. Kurien deserve mention in this context.

Reforms that took decades, if not centuries, elsewhere in the world were brought in, literally overnight, with a universal franchise (including votes for women) being a striking example. The country also secured what was practically an instantaneous release, from social and economic evils such as dowry, child marriage, sati (the ancient Hindu practice of the widow jumping into the funeral pyre following the husband's death) and untouchability. The establishment of robust and vibrant constitutional institutions such as the Supreme Court of India, the Election Commission and the Union Public Service Commission, all of which have acquitted themselves admirably despite many trials and tribulations, has added immense respectability to the quality of democratic governance. The setting up of Centres of Excellence such as the Indian Institutes of Technology (IITs) and Indian Institutes Management (IIMs) has also won laurels for India, with their alumni being sought after the world over. Citizens of India have covered themselves with glory in various fields of human endeavour - from sports, culture, and exploration to literature and economics, to scaling the dizzy heights of the corporate world. Names such as those of Rabindranath Tagore, Sir C.V. Raman, Sachin Tendulkar, P.V. Sindhu, Vishwanathan Anand, Satya Nadella and Amitabh Bachchan, spring to mind in this context.

Doctors and engineers of Indian origin have established themselves in every nook and cranny of the world. Indian doctors are regarded as among the best wherever they are, and Indian engineers have made substantial contributions to the Information Technology (IT) sector to the National Aeronautical and Space Administration (NASA) in the United States of America (USA). Indian science has also come of age - Antarctica has been explored and so have the Moon and planet Venus. Thanks to the structure of the polity it has adopted, including the system of checks and balances and separation of powers -as between the Judiciary, the Executive and the Legislature, and with the Election Commission being able to conduct free, fair, and impartial elections regularly, the country has not only sustained the democratic form of government but has made it flourish and bloom. And the two pillars that constitute the foundation of a safe democratic polity, as recognised by philosopher Voltaire, namely universal literacy and a free media, can also be safely said to have been ensured by the system.

The manner in which the Reserve Bank of India (RBI) managed to run a tight ship, and maintain course, in terms of maintaining a firm grip over the economy, at the time of the financial meltdown that, in 2008, shook the economic foundations of the world – was acclaimed even by the Chairman of the Federal Bank of the United States at that time. The country has also been second to none in espousing, and joining of hands with the comity of nations, in addressing major challenges facing humanity such as atmospheric pollution, environmental degradation, climate change, and terrorism. Even in addressing relatively recent global concerns such as legislating to introduce the Right to Information Act (CRTI) putting in place mechanisms for promoting and encouraging Corporate Social Responsibility, steps taken by India have received universal acclaim.

If the culmination of the Independence Movement, in the grant of freedom, was the first major watershed in the history of modern India, the advent of the forces of liberalisation, privatisation and globalisation, and the structural adjustment that the Indian economy had to make in their wake, surely represented on second such important event. It is indeed a matter for gratification that the country has managed, not merely to survive the onslaught of a set of hitherto unfamiliar forces, but to emerge the stronger for having overcome their challenge.

However, it is not as though the picture is entirely rosy. Hunger, poverty exclusion, fear, and deprivation are still writ large in pockets across the country and continue to plague some sections of the population. Scourges such as honour killings of women, trading in children and acute distress in the agriculture sector, leading to the ugly spectre of suicides by farmers, are all things that hardly redound to the credit of the country's image. It is indeed a paradox that a country that is home to the Tatas, the Birlas and the Ambanis should also be the one which is at a pitiable 129th rank in 2019, Human Development Index (HDI) ranking of the United Nations Development Programme (UNDP).

In other words, as one pauses for breath, sits back and looks at the big picture, one finds that the story of independent India presents a mixed bag indeed. A great deal has been achieved. But there is much that remains to be done. There is cause neither for despair nor undue complacency. The residual agenda is very much doable, given determination, hard work, the ability to address the items on the national agenda in order of priority, and the will to succeed.

The "to do" list is formidable, but not overwhelming. However, the fact that certain objectives constitute the bottomline of the nation's agenda is beyond controversy. These include defending and protecting the country's sovereignty, securing sustainable and equitable growth of the economy, and maintaining the levels of food security. And doing all that in the face of looming threats such as food insecurity, climate change, terrorism and pandemics. Clearly, the need to achieve equitable and steady growth of the agriculture and allied sectors will constitute an overriding priority in that effort.

The scourge of farmers ending their lives on account of economic distress has for long engaged the attention of various fora across the country for some time now. Emotional and acrimonious debates are being witnessed in Parliament and the Legislatures of various states. Given that the tragic and unacceptable phenomenon of farmers committing suicides continues unabated, and keeping in mind the tremendous capacity and reach of the State and the Central governments, it would appear that either their understanding of what is possible or their plan of action is entirely off-target.

The relationship of India's agriculture sector to the economy has undergone a significant transformation in recent years, with its share of the GDP falling, steadily from 50% in 1960 to about 17% at current prices (2019-20), indicating a shift away from the traditional agrarian character of the economy towards one dominated by the services sector. This decrease has not, however, been accompanied by a matching reduction in the share of agriculture in employment. More than 54.6%of the total workforce is still employed in Agriculture and Allied Sectors and a large proportion of the population is dependent on agriculture for sustenance.

Employment in agriculture (% of total employment) (modeled ILO estimate) in India was reported at 41.49 % in 2020, according to the World Bank collection of development indicators, compiled from officially recognized sources. India - Employment in agriculture (% of total employment) - actual values, historical data, forecasts and projections were sourced from the World Bank on October of 2020.

Employment in Agriculture (%) of total employment

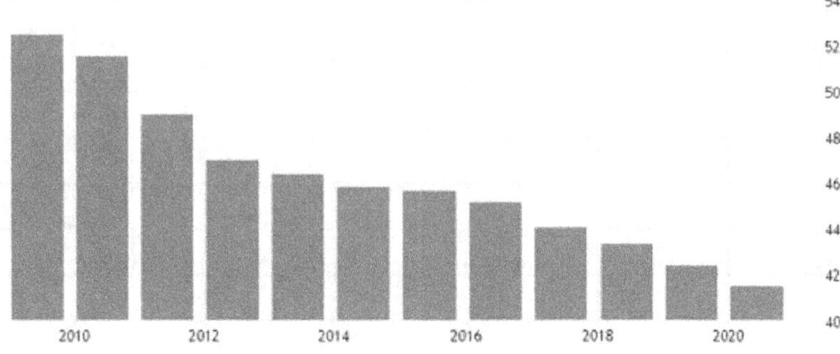

Source: World Bank, 2020

On the other hand, as against a target of 4 per cent for agriculture and allied sectors in the Twelfth Plan (2012-17), the growth registered in the first year (at 2011- 12 prices) was 1.2 per cent in 2012-13, 3.7 per cent in 2013- 14, and 1.1 per cent in 2014-15. Gross capital formation in agriculture as a percentage of GCF (Gross Capital Formation) which was 18.23% in 2011, felt to 13.82% in 2016-17 at current prices 2011-12 (source: State of Indian Agriculture 2017, GoI) GDP, which was 20.69% in 1993-94, fell to 4.99% in 2011-12 at constant (2004-05) prices. Even in a year of record growth (2006-07), the contribution of the sector actually fell. The state of the sector is, indeed, a matter of grave concern – especially in the context of the size of the population and the ever-increasing threat of looming food security.

Percentage of people employed in agriculture has been consistently declining, from around 60% in 1999-00 to 49% in 2011-12. In absolute terms, between 2004-05 and 2011-12, there has been a net reduction of 30.57 million of labour from the agricultural sector.

Food security cannot be de-linked from the need to maintain the viability of farm holdings. Increased production and enhanced productivity do not necessarily reflect healthy farming systems. As a matter of fact, the majority of suicides by farmers' happened in the wake of gluts in commodities such as cotton, tobacco, chillies, potato, and onion, following record production and yields. Growth can be achieved and sustained, only if farm incomes rise and the farmers' share in the consumer-rupee increases, through the realisation of remunerative prices and reduction of transaction costs.

The country's public agriculture management systems, in their present approach, appear to have grabbed the wrong end of the stick: with the mandate of the (national- and state-level) agriculture departments remaining the measurement of production and enhanced productivity rather than the size of farm incomes. The country comprises various agro-ecological situations (five, fifteen, or a hundred and twenty, depending on the degree of detail employed). These maintain a dynamic and sensitive relationship with the external environment - which comprises factors such as agro-climatic conditions and the performance of support systems providing backward and forward linkages in the value-chain – such as irrigation, power, research, extension, technology transfer, credit, insurance, supply of farm inputs-seeds, fertilizers and pesticides - and storage/processing/marketing. Although agriculture is a state subject, in the arrangements stipulated by the provisions of the Constitution of India, the central government remains the source for the funding of most of the programmes currently under implementation. This fact, and the largely top-down, rigid and straitjacketed guidelines that inform the design of these programmes have, overtime, led to a lack of response to regionally differentiated sectorial/sectional concerns; and resulted in the states remaining passive recipients of central assistance.

A quick and close look at the performance of the agriculture sector reveals the first major concern. The important players among those that support the value chain of agriculture remain - chronically, it could be said - in a supply mode. Put another way, they practise the 'cafeteria' approach: a somewhat unilaterally prepared menu is served up and the farmer is required to choose only from the items available therein. The organisations involved have, thus far, not exhibited the ability, or the will, to tailor products capable of responding to variations in the pattern and quality of demand, in a demand-driven regime in which varying differences in demand, across crops, regions and farmer – sections, are primary considerations that inform the product design.

A paradigm shift, therefore, is urgently needed, away from the current supply-dominated mode to a demand-driven regime – not only in respect of the design and architecture of the programmes of the central government, but also in terms of the pattern and quality of the response of the support systems, to the demands of the sector.

A sharper definition, and recognition, of the allocation of roles between the central and the state governments, is an urgent need. The central government's preoccupation with running a large number of programmes has also led to a dangerous disconnect with the outside world with little meaningful dialogue taking place with the rapidly changing external environment – an overriding imperative, especially post-liberalisation, privatisation and globalisation.

The Ministry of Agriculture, GOI needs to undergo a self-imposed "one-time-catch up" exercise to remain *au fait* with the rapidly changing external environment. So should the players in the value chain. In addition, there is also no effort at macro-management in important areas of the centre's basic functions - such as credit, insurance, research and Exim Policy (which is currently, at best, stop-go and knee-jerk in nature). Many important and urgent steps need to be taken to ensure the rapid and sustainable growth of the agriculture and allied sectors in the future, and for farm-incomes to increase to expected levels.

Firstly, it must be appreciated that the process of decision-making in agriculture and related sectors occurs in a complex and vast administrative space. Therefore, the requisite speed and accuracy cannot obtain unless the highest levels in government at the Centre and state levels directly participate in the process and lead from the front. To that end, it is necessary to constitute a Cabinet Committee presided over by the Prime Minister at the national level in which Ministers responsible for all the departments concerned with agriculture and the allied sectors are Members. The Committee should be serviced by the National Institute of Agricultural Extension Management (MANAGE) through a trans-disciplinary team (whose functioning is informed by a process of chemistry among its members rather than mere physics). Disciplines such as research, extension, meteorology, financial services (including banking and insurance), agronomy, animal husbandry, storage, processing and marketing. Similar mechanisms should function at the state and district levels with the Chief Ministers and the District Magistrates/Collectors, respectively as the heads – serviced by State Agricultural Extension Management and Training Institute (SAMETI) and Agricultural Technology Management Agency (ATMA), respectively. At all levels, an online and real-time scanning should be undertaken with a view to spotting opportunities and threats, adding value thereto and dispatching the resultant messages to appropriate destinations in the country.

The centre and the states should function on their own in certain spheres. For instance, the Centre alone will need to deal with areas such as international cooperation, international trade and financial services such as banking and insurance. The centre and states should act in a complementary manner in areas such as the setting up of and maintenance of infrastructure such as ports, railways, national highways, cold storage chains and the like, and management of the National Research System in the agriculture and allied sectors. Similarly, the states should deal exclusively with areas such as extension, the designing, preparation, and implementation of district plans to be financed by the funds available at the state level, etc. In all other areas the Centre and the states should supplement the activities of one another.

In place of the extant arrangements, the centre and the states should jointly enter into Memoranda of Understanding (MoU)s which promise pre-set, and quantified deliverable outputs in terms of production, productivity, and other thematic imperatives such as soil health, centre-staging the role of women, NRM, environment-friendly practices, robust responses to the challenges of climate change, etc. The sum of the outputs and outcomes of all the MoUs in the country should, ideally, translate into national objectives, both in terms of quantities and thematic imperatives. The RKVY (Rashtriya Krishi Vikas

Yojana) – RAAFTAR (Remunerative Approaches to Agriculture and Allied Sectors Rejuvenation) programme of the Government of India (GoI) has made a good beginning. It however, needs to be taken forward purposefully.

This book is addressed both to policy makers, professionals and researchers academicians, and extension personnel working in the field of agricultural development and policy. The objective is to provide relevant theoretical frameworks and the latest empirical research findings in agricultural development and policy.

Mohan Kanda

Acknowledgements

As is always the case, the task of writing this book was the outcome of a dedicated band of workers comprising a team of which I was the leader who, at best, dotted the 'i's crossed the 't's of the output of the group.

Nearly two full years of unrelenting hard work has gone into this effort, the main contributors being, not necessarily in the order of the importance of their contribution, Prof. J. Devi Prasad, Director and his team K. Padmaja, M. Raja Krishna Murthy from the Centre for Good Governance, Hyderabad, and my Man K.V. Ramakrishna Rao, P.V.A. Ramarao, who has been my friend, philosopher and guide for over a decade now, holding my hand and guiding me through the complicated, and, on occasion, somewhat frustrating task of beginning and completing many books, (on such varied subjects as Disaster Management, Agriculture and Ethics in Governance), has made a substantial contribution to the content and value of the book.

My colleagues in the civil service, Atul Sinha and Suresh Kumar, made several valuable suggestions to improve the format and presentation of the book. G. Jayalakshmi of Director General, National Institute of Agricultural Extension Management (MANAGE), Hyderabad took the trouble of getting the various chapters vetted by her colleagues who responded with spontaneous warmth and promptness.

Prof. T.V.K. Singh, Dean of Agriculture, PJTSAU (Retd.) and ANGRAU has done an outstanding job of ensuring that the information present in the book is up-to-date and accurate. I am thankful to him.

Late Dr. S.M. Hassan, Joint Director of Agriculture (Retd.), Consultant and P. Ashok Kumar, Junior Technical Assistant, Technical Support Group, Agriculture and Cooperation Department, Government of Telangana have, once again, made an enormous contribution by way of repeated directions of the proofs at the various stages and its final shape is largely on account of his contribution.

My nephew Raghu Kanda and the two young ladies from Prerana Foundation, Ranjeeta Bhat and K. Vani Praveena, spent several hours patiently thumbing through the pages and correcting errors in sequencing, grammar and syntax.

Anil Shah has once more proved to be a publisher *par excellence* in the manner in which the get up of this book was finalised.

(xvi)   Acknowledgements

I am grateful to Sri. K. Padmanabhaiah, (formerly Secretary, Ministry of Home Affairs, Government of India), Chairman, Board of Governors, Administrative College of India, Dr. Trilochan Mohapatra, Director General, ICAR, Government of India and Smt. Vasudha Mishra, Secretary, Union Public Service Commission, Government of India who readily agreed to contribute blurbs.

I am indebted to Prof. Ramesh Chand, Member NITI Aayog and Member 15th Finance Commission for taking time out to write a foreword to the book.

Mohan Kanda

Contents

CHAPTER – 3: SUSTAINABLE GROWTH AND EQUITABLE DEVELOPMENT

CHAPTER – 4: THE ROAD AHEAD

CHAPTER - 1

AGRICULTURE IN INDIA

"If Agriculture goes wrong, nothing else will go right in the country"
- M.S. Swaminathan

India experienced impressive growth and productivity gains in agriculture since Independence reflecting our enterprising farmers' resilience against multiple odds and challenges. Despite a structural transformation through green, yellow, white and blue revolutions characterised by food security policy objective, the regular distress and crises in the recent past pose a severe threat to the income and livelihoods security of farmers. Farmers are at the epicentre of the Indian economy, and their livelihood upliftment is a step towards the holistic development of the nation. Therefore, enhancing the incomes of farmers and ensuring their income security, thus, has been of concern to all. Unless farmers' income increases substantially, distress cannot be tackled.

Some possible options for enhancing farmers' income are

1. Enhancing Gross Income
2. Increasing Production
3. Enhancing Output Price
4. Diversifying within the farm sector
5. Diversification to non-farm sector
6. Reducing costs
7. Stabilising income and risk mitigation
8. Expanding irrigation
9. Diversification to High-value crops/Enterprises
10. Better price realisation
11. Improving terms of trade for agriculture
12. Technology up-gradation
13. Non-farm activities bolstering the livestock sector
14. Tapping solar power on farmers' fields and so on.

The major constraints, for doubling of farmers' income by 2022 within 5 years (i.e., 2018-2022) are as follows:

1. Low and unrealisable Minimum Support Price (MSP)
2. Non-remunerative price in the market
3. Low share of farmers in final price
4. Poor penetration of crop insurance
5. High and increasing input cost and
6. Absence of adequate and required market infrastructure, etc.

This chapter, set in this backdrop, says that doubling of farmers' income is possible as there exists potential and explores different options for enhancing farmers' incomes while elaborating a few important of them. We conduct a quick recap of the agriculture revolutions, the past trends in farmers' incomes, the present situation, government interventions, research, extension, metrological services, financial services including banking, credit and insurance, input (seeds, implements, fertilizers, and pesticides), storage, food processing, and marketing, respectively. Further, we discuss certain areas where we need to focus more and differently.

1.1 A Quick Recap

Although India is amongst the world's leading food producers in the world, low yields persist despite the Green Revolution. It is the second-largest producer of rice and wheat in the world but, in terms of yield, it is ranked much below. The agriculture sector also suffers from issues such as high dependency on the monsoons, a weak post-harvest infrastructure, underdeveloped agro processing facilities and high levels of post-harvest losses, which, together, adversely impact productivity levels. This is one reason why the contribution of the sector has declined a mere 15.87 per cent of the national output, in spite of the sizable proportion of people depending on it still remaining at about 58.0 per cent.

The income levels of farmers in India have remained significantly low. The National Sample Survey Office (NSSO) survey results show that in 2011-12, nearly 23 per cent of the farm households earned incomes less than the poverty line, with this proportion being much higher in states like Jharkhand (45 per cent), Odisha (32.1 per cent), Bihar (28 per cent) and Madhya Pradesh (26 per cent). Income levels have also been unstable, as they have mainly followed the pattern of food production which has fluctuated over the years primarily on account of the changing climate which has made rainfall erratic and unpredictable in recent years, leading to a substantial increase incrop failures. This has led to agrarian distress and agitations. Farm loan waivers have never been useful to alleviate distress; long-term solutions are needed to address the deeper structural problems which continue to plague the sector.

- *Weak post-harvest infrastructure and high food wastage*: About 35-40 per cent of the total food production in India is wasted every year. Part of the wastage takes place at the farm level due to pests, weeds, other diseases and lack of proper storage facilities.

- *Low returns on farm produce*: Inadequate infrastructure facilities coupled with poor market linkages has led to the creation of multiple levels of aggregators/intermediaries in India who take care of transportation and distribution of food. With a mark-up to producer price at each aggregate or interface, the gap between the farm gate price and consumer price widens, with farmers receiving much lower prices for their produce as compared to the price at which they are sold in the terminal retail markets.

Farmers also suffer on account of the cobweb phenomenon. It has been observed in India that whenever prices of a commodity increase during a season of scarcity, farmers tend to increase the cultivation of the same commodity, leading to a problem of plenty, which consequently leads to a decline in its prices, causing a huge loss to farmers. That, in turn, makes farmers turn away from that commodity in the succeeding season.

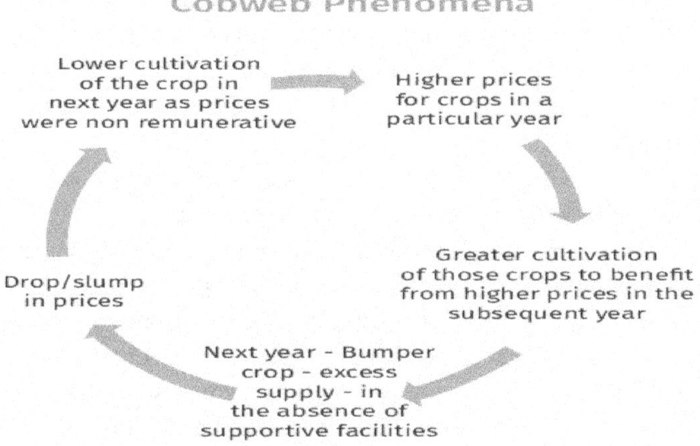

Figure 1.1 Cobweb Phenomenon

Other factors responsible for affecting the returns of farm produce are:

1. Weakness in the APMC Act limit the wholesale prices received by farmers: Traders in Agricultural Produce Market Committees (APMCs) have often been found to be forming cartels and obstructing transparent price discovery. Small farmers are offered lower prices while the same

items are sold to large buyers such as rice/flour mill owners at higher prices. The license fee in APMCs is also prohibitive very high and restricts the entry of new buyers, resulting in the continuation of the monopoly of existing traders and their cartels.

2. Restrictive trade policies prevent exporters from tapping export opportunities - Continued restrictions on exports during times of excess production lead to a fall in farmer incomes. Policy interventions have reportedly reduced gross farm revenues by over six per cent per year between 2014 and 2016.

   The above issues have been addressed by the GoI recently in the following ordinances

   (i) The Farmers (Empowerment and Protection) Agreement on Price Assurance and Farm Services Ordinance, 2020

   (ii) The Farmers Produce Trade and Commerce (Promotion and facilitation) Ordinance, 2020 under clause (1) of article 123 of the Constitution of India

3. Overstocking of food grains by the Food Corporation of India locks a substantial quantity that can be made available for exports and can fetch higher returns for farmers.

4. Input subsidies (power and fertilisers) have led to deteriorating soil quality and depletion of ground water far in excess of the utilisable recharge, resulting in alarming falls in water tables across the country.

5. While the introduction of the Soil Health Card Scheme has had a positive impact, there are concerns regarding inadequate infrastructure (laboratories and manpower).

6. Drought proofing measures such as micro-irrigation have not succeeded and have had little impact on water availability.

7. Minimum Support Price (MSP) measure taken to insure farmers against price crashes and ensure food security, has not been fully effective. The low level of awareness of the MSP scheme and procurement practices has hindered the effectiveness of the scheme. In addition, limited coverage of crops (23) under MSP and providing heavy subsidies for inputs such as fertilisers, seeds, and pesticides for certain crops have led to a skewed food basket and uneconomic cropping pattern, leading to large surpluses in some crops and shortages in others. Domestic policies (MSPs) may also encourage the production of locally consumed commodities and, thus, farmers fail to capture available export opportunities for commercial and horticulture crops.

8. Farm loan waivers have provided no relief and have only vitiated the repayment climate apart from reducing the size of the overall kitty available for lending.

9. Frequent changes in trade policy, which are used largely as a price stabilisation tool to provide necessary price support to farmers have restricted export growth.

10. Heavy subsidies provided for inputs such as fertilisers, seeds, and pesticides are mostly accessed only by large farmers.

Historical Developments

Great Bengal Famine of 1770

The Great Bengal Famine of 1770 occurred between 1769 to 1770, affecting the lower Gangetic plain of India from Bihar to the Bengal region. It is estimated to have claimed 10 million lives. The report of Warren Hastings, the then Governor General of India, estimated that a third of the population in the affected region starved to death. It was one of the many famines and famine-triggered epidemics that devastated the Indian subcontinent during the 18th and 19th centuries. It is usually attributed to a combination of climatic conditions and the policies of the British East India Company. The start of the famine has been attributed to a failed monsoon in 1769 that caused widespread drought and two consecutive failed paddy crops. The poor infrastructure investments in the pre-British period, devastation from war, and exploitative tax revenue maximization policies of the British East India Company after 1765 crippled the economic resources of the rural population. Nobel Prize winning Indian economist Amartya Sen describes it as a man-made famine, noting that no previous famine had occurred in Bengal that century.

As a result of the famine, large areas were depopulated and turned to jungles for decades to come, as the survivors migrated in search of food. Many cultivated lands were abandoned-much of Birbhum, for instance, returned to jungle and was virtually impossible for decades afterwards. From 1772 onwards, bands of bandits and Thugs became an established feature of Bengal, and were only brought under control by punitive actions in the 1890s.

The most common approach to famines is to propose explanations in terms of the Food Availability Decline (FAD). This FAD approach has been extensively used to analyse and explain the Bengal famine. The Famine Inquiry Commission's view that the primary cause of the famine was 'a serious shortage in the total supply of rice available for consumption in Bengal' provides the standard explanation of the famine, as George Blyn notes in his authoritative account of 'agricultural trends in India' (1966), referring to the Report of the Famine Inquiry Commission and to the Census of India 1951. In 1942-43 cyclones and floods reduced the Bengal rice crop by about a third; this, coupled with the absence of exports from Japanese-controlled Burma, and inadequate relief, led to famines, epidemics (malaria, cholera and smallpox), aggravated by widespread starvation'.

Green Revolution 1960s

The Green Revolution, or the Third Agricultural Revolution, is a set of research and technology transfer initiatives occurring between 1950 and the late 1960s that increased agricultural production worldwide, particularly in the developing world, beginning most markedly in the late 1960s. The initiatives resulted in the adoption of new technologies, including high-yielding varieties (HYVs) of cereals, especially dwarf wheat and rice, in association with chemical fertilizers and agro-chemicals, and with controlled water supply (usually involving irrigation) and new methods of cultivation, including mechanization. All of these together were seen as a 'package of practices' to supersede 'traditional' technology and to be adopted as a whole.

Both the Ford Foundation and the Rockfeller Foundation were heavily involved. One key leader was Norman Borlaug, the "Father of the Green Revolution", who received the Nobel Peace Prize in 1970. He is credited with saving over a billion people from starvation. The basic approach was the development of high-yielding varieties of cereal grains, expansion of irrigation infrastructure, modernization of management techniques, distribution of hybridized seeds, synthetic fertilizers, and pesticides to farmers.

The term "Green Revolution" was first used in a speech by the administrator of the U.S. Agency for International Development (USAID), William S. Gaud, who noted the spread of the new technologies: "These and other developments in the field of agriculture contain the makings of a new revolution. It is not a violent Red Revolution like that of the Soviets, nor is it a White Revolution like that of the Shah of Iran. I call it the Green Revolution."

In 1961, India was on the brink of mass famine. Dr. Norman Borlaug was invited to India by the adviser to the Indian Minister of Agriculture, Dr. M.S. Swaminathan. Despite bureaucratic hurdles imposed by India's grain monopolies, the Ford Foundation and Indian Government collaborated to import wheat seed from the International Maize and Wheat Improvement Centre (CIMMYT). Punjab was selected by the Indian government to be the first site to try the new crops because of its reliable water supply and a history of agricultural success. India began its own Green Revolution program of plant breeding, irrigation development, and financing of agrochemicals.

The two successive severe droughts in 1965-66 and 1966-67, gave rise to international apprehensions about India's capacity to feed her huge and growing population. The harshest critics recommended the application of the "triage" formula to countries like India which were considered beyond redemption.

Fortunately for the country, at this very time the High Yielding Varieties (HYVs) of cereals became commercially available. India's policymakers plumped for it with alacrity. Dr. Norman Borlaug complimented the then Minister C. Subramaniam, a visionary for Agriculture as "the first high officer to recognize the significance of the new wheat strains and willing to take the risk involved in importing 18,000 tonnes of dwarf Mexican varieties".

The Pearson Report characterized the speedy adoption of HYVs as "one of the authentic marvels of our time". Others described the process of agricultural transformation as "one of the most amazing stories of our time". While this was ageneral observation, the economists, who had neither anticipated the Green Revolution nor played any part in its adoption by way of even policy advice, did not take kindly to it. Their reaction varied from scepticism ("Cornucopia or Pandora's Box") to downright condemnation on the ground that it was leading towards the emergence of dualism. Let us accept that technological changes ushered through the application of HYVs "as such have contributed to the widening of the income disparities between 1. different regions, 2. small and large farms and 3. land owners on the one hand and tenants and agricultural labourers on the other". But the question is: situated as the country was in the mid-Sixties, when its capacity to feed its people was seriously being questioned, and some critics were advocating the application of "triage" and "life boats" formula to food aid, what was the choice before the policymaker? The highest priority had to be assigned to augmenting food production and the HYVs offered an excellent means of doing so. The possibility of its in egalitarian effects - assuming that these could be clearly perceived at that time - had to be weighed against the obvious in egalitarian effects of food shortage and high prices, under which the poor suffer the most.

It is also asserted that "despite technological changes, the growth of agricultural output in India slowed down in the 1960s compared to 1950s".

Post-Independence Growth of Indian Agriculture

Indian agriculture has come a long way since the inception of planning in 1951. All along there was an almost obsessive concern of development policy with the attainment of self-sufficiency in food. The country faced an acute shortage of food grains in the 1960s, when the average annual imports of wheat hovered around three million tonnes. It survived through this phase because of the munificence of the US, which supplied wheat to us under its PL-480 programme. This donor-donee relationship was far from flattering, as revealed by M.L Dantwala in the following quotation, reproduced, from the book 'Famine' 1975 by Paddock Brothers, reflecting the perception of the US: "America will have to apply classical medical 'triage' method. Like doctors on the battlefield trying to make the best out of minimum resources, will have to decide which countries to save and which to sacrifice. Today India absorbs like a blotter 25 percent of the entire American wheat crop. No matter how one may adjust present statistics and allow for future increases in the American wheat crop, it will be beyond the resources of the US to keep famine out of India during the 1970s. Of all national leaderships, the Indian comes close to being the most childish and inefficient and perversely determined to cut the country's economic throat." The moral: If other more deserving countries are to be saved, India must be sacrificed. At the time of our gaining independence, the first Prime Minister Jawaharlal Nehru said, "everything else can wait, but not agriculture." There have been several policy statements for agriculture during the last sixty years. Thanks to the Green Revolution, India attained self-sufficiency in food grains in the 1970s and, what is more, emerged as an exporter of food grains in recent years.

Attainment of food self-sufficiency, even in a technical sense, is admittedly an important landmark in the history of Indian agriculture development, but this has not meant the dilution of problems. Indian agriculture is facing problems of agricultural growth and rural development in general continue to remain, in the new millennium, as daunting as they were in the 1950s. One major difference is that the agricultural sector today faces a host of what we might call second generation problems on which we would seek to focus.

The post-Independence development of Indian agriculture can be broadly grouped into four phases. The first phase (1947-64) was the Nehruvian era where the major emphasis was on the development of infrastructure for scientific agriculture and a large expansion of area under irrigation. During this period, the population started increasing by over 3 percent a year as a result of both the steps taken to strengthen public health care systems and advances in preventive and curative medicine. The growth in food production was inadequate to meet the consumption needs of the growing population, and food imports became essential. Such food imports, largely under the PL-480 programme of the United States, touched a peak of 10 million tonnes in 1966, (M.S. Swaminathan, 2003). In this period, there was an interesting debate between structuralist' school, i.e., those believing in drastic changes in land relations as a prerequisite for effecting a breakthrough in agriculture and those who thought, based on the evidence of high output response to irrigation and new technology, that high agricultural growth was possible through input intensification, despite the smallness of farm size and prevalence of tenancy. It became clear that the potential for agricultural growth from the investments made in irrigation, and from the autonomous factors, e.g., the expansion of area under cultivation, intensives created by the implementation of land reforms like the abolition of intermediaries and the rise of agricultural classes to political power were nearly exhausted, and that further growth of agriculture depended crucially on the expansion of agricultural infrastructure and the application of new technology to raise farm productivity and profitability.

The large imports of food grains in the wake of two-successive droughts in mid-sixties, and the unacceptably high political costs that it entailed, pushed the government towards a bold strategy for achieving self-sufficiency in food grains through the green revolution by stepping up investments in irrigation, evolving and applying High-Yielding Varieties of seeds and intensifying the use of inputs like fertilizers. The results of the adoption of the High Yielding Variety (HYV) programme were quick and substantial. Food grains production which had hovered around 50.82 million tonnes per year since 1950-51 started increasing at a fast rate.

In the second phase (1965-1985), India underwent a radical change in the production of food grains from the mid-60s onwards, consequently, "in 1968, several thousand year old barrier in the yield of wheat was broken and India achieved a wheat production of 17 million tonnes. An American scientist Dr.William Gaud called the dramatic breakthrough the "Green Revolution". The main achievement was in the area of wheat production and therefore many economists called it the Wheat Revolution instead of the Green Revolution. The advent of the Green Revolution was at a time when the availability of additional land had more or less reached its limits, the agricultural scenario changed from one of land reclamations to one heavily dependent on modern inputs.

The introduction and rapid spread of High Yielding rice and wheat varieties resulted in steady output growth for food grains. The production which was 10.40 million tonnes in 1965- 66, rose to 99.70 million tonnes in 2017-18. Wheat exports which were 1.64 Lakh tonnes 2009-10 with a value of 231.90 crores have now come down to a mere 0.11 Lakh tonnes with a value of 26.92 crores. Public investment in irrigation and other rural infrastructure, research and extension together with improved crop production practices has significantly helped to expand production and stock to food grains. This increase in food grains production has helped the country to achieve a considerable degree of self-sufficiency in terms of food requirements and tide over recurring food shortages reminiscent of the 1960s and 1970s.

The third Phase (1985-2000) was characterized by the greater emphasis on the production of pulses and oilseeds as well as of vegetables, fruits, and milk. This period ended with large grain reserves with the government, with the media highlighting the co-existence of "Grain Mountains and hungry millions".

Indian Agriculture during the 1980s

The poor growth record of the early Green Revolution period was reversed after 1980-81, a period we call the "late Green Revolution period". As Bhalla and Singh (2001) noted, the 1980s represent a period of the spread of the Green

Revolution to larger areas and more crops. Some authors have called the 1980s as the phase of "wider technology dissemination" (Chand, 2004). Food grain yields increased at an annual rate of 3.2 percent between 1981-82 and 1991-92. A sharp increase in rice yields accounted for most of the increase in food grain production, rice yields grew annually at 3.3 per cent between 1981-82 and 1991-92 compared to 1.5 per cent between 1967-68 and 1980-81. The agriculture GDP also registered an impressive annual growth rate of 3.4 per cent in the 1980s. There were two important factors that contributed to the turnaround in the 1980s: first, there was a major jump in production in the eastern region of the country, particularly in the State of West Bengal. Secondly, there was a major improvement in the production of oilseeds in the central Indian region.

Indian Agriculture during the 1990s

By the late-eighties and the early-nineties, the official policy on agriculture followed until then came to be criticized. This critique of the earlier policy was led by a section of economists as well as international financial institutions, such as the World Bank, all wedded to the ideas of the Washington Consensus. It was argued that the earlier policy deliberately skewed the terms of trade against agriculture through protectionist industrial and trade policies and an overvalued exchange rate. It was argued that once we "get the prices right", the incentive structure in agriculture would improve, and farmers would respond to higher prices by producing more.

According to one study if domestic prices had been aligned with world prices, average incomes in agriculture in the early-1990s would have been 16 to 25 per cent higher than what they actually were. Liberalization of agricultural trade was an important step in imparting efficiency to Indian agriculture.

These arguments were derived primarily from the standpoint of the neo-classical trade theory, in which free trade and openness would maximize efficiency and gains. It was argued that India has major comparative advantages in diversifying its cropping pattern in favours of high value, export-oriented crops like fruits, vegetables and flowers. Further, restrictions on private stocks and internal trade should be eliminated, which would help to evolve a national market in agriculture. The argument of the proponents of the new policy has been that once terms of trade improve; price incentives would generate a significant supply response. However, the vast literature on the supply responsiveness of farmers has shown that the relationship between prices and output is very weak. There are, of course, major issues related to the accuracy of economic models used to estimate supply response in agriculture, such as the measurement and control of different effects. Yet, the range of long run supply elasticity of aggregate agricultural output has historically been between 0.1 and 0.5 in developing countries.

India has 198.36 million hectares of cultivable land with 157.81million agricultural holdings. The average operational holding is1.08 hectares, and about two-thirds of the cultivable land solely dependent on monsoons. Indian agriculturists with all the constraints have to compete with farmers from the rest of the world with operational holdings exceeding 1,000 hectares. Indian agriculture has to be modernized to compete in the world. The productivity of crops has to improve through the introduction of micro-irrigation systems and new farming technology, strengthening extension services, improved post-harvest management and bio-technology, promoting information technology and enhancing market leadership. There had been some improvement, particularly during the Five Year Plan periods. During the year 2005-06, actual percentage Growth in Agriculture and allied sector was 4.9 percent over three years (2005-06 to 2007-08), it has reduced to 1.6 in the year 2008-09.

The package of macro-economic and trade policy reforms introduced in 1991 consisted of macro-economic policy changes, changes in exchange and trade policy, devaluation of the currency, gradual dismantling of the industrial licensing system and controls, reduction of tariffs, reform of public enterprises and increasing privatization. Although no direct reference was made to agriculture, the new policy framework was expected to be highly beneficial to tradable agriculture through ending discrimination against it.

Table 1.1 Average GDP Growth Rates in Agriculture
(% per Year at 1999–2000 Price)

S.No.	Period	Total Economy	Agriculture and Allied Sectors	Crops and Livestock
1	Wider technology Dissemination period 1981–82 to 1990–91	5.40	3.52	3.65
2	Early reforms period 1991–1992 to 1996–97	5.69	3.66	3.68
3	Ninth Plan 1997–98 to 2001–02	5.52	2.50	2.49
4	Tenth Plan period 2002–03 to 2004–05	7.77	2.47	2.51
	to 2006–07 of which 2002–03	6.60	0.89	0.89
	2005–06 to 2006–07	9.51	4.84	4.96

Source: National Accounts Statistics 2008 (New Series), Central Statistical Organization, Ministry of Statistics and Programme Implementation, New Delhi.

The liberalizers argued that the import substitution strategy of industrialization under the planning regime followed by most developing countries in the post war period was highly discriminatory against the agricultural sector. But despite the changes in the macro-economic policy framework and trade liberalization, India's agricultural sector did not experience any significant growth subsequent to the initiation of economic reforms in 1991. In fact, except for a short period 1991-92 to 1996-97, when because of a highly favourable international climate, agricultural exports rose sharply, the agricultural sector has not derived the expected benefits from trade liberalization. Nor has the new macro-economic policy framework resulted in accelerating agricultural growth. In fact, when compared with the immediate pre-liberalization period 1980-81 to 1990-91, agricultural growth in India recorded a visible deceleration during the post-liberalization period 1990-93 to 2003-06.

Agriculture during the 2000s

In the fourth phase, a further increase in food grains production became difficult. According to the GoI's Economic Survey, 1999-2000, "There are limits to increasing production through area expansion as the country has almost reached a plateau in so far as cultivable land is concerned. Hence, the emphasis has to be on increasing productivity levels. The area under food grains has more or less remained constant at around 125 million hectares since 1970-71.

The agricultural decline is taking place at a time when international prices of major food grains are going up steeply, partly owing to the use of grain for ethanol production. Land for food versus fuel is becoming a major issue. For example, the export price of wheat has risen from $197 a ton in 2005 to $263 a ton in 2007. Maize price has gone up from about $100 a ton in 2005 to $166 a ton now. International trade is also becoming free but not fair. There is also an indication of adverse changes in rainfall, temperature, and sea level as a result of global warming.

While the government rejoices over a record total food grain production in the country is estimated at record 291.95 million tonnes which is higher by 6.74 million tonnes than the production of food grain of 285.21 million tonnes achieved during 2018-19. There are doubts about the country's ability to produce enough to meet demand by 2033-34, if agricultural production does not remain above the population growth rate. In India there is a need to increase the annual food grain production from the present 295.67 million tonnes to 340-355 million tonnes by 2033-34 based on actual consumption, i.e., 2033-34 as estimated by NITI Aayog working group (February 2018). Our agriculture is at the crossroads economically, ecologically, technologically, socially and nutritionally. A "business as usual approach" in the farm sector now will lead to an unprecedented human calamity, the beginnings of which we are now witnessing in the form of suicides by farmers in several parts of the country, including the Punjab which is the heartland of intensive agriculture, (Government of India, 2004).

Structural Adjustment of India's Economy 1991-2010

In 1991, India faced an unprecedented balance of payments crisis. For almost a decade the government had borrowed heavily to support an economic strategy that relied on expansionary public spending to finance growth. From 1980 to 1991 India's domestic public debt increased steadily, from 36 percent to 56 percent of the GDP, while its external debt more than tripled in 2013-14to $70 billion.

Political changes, unrest in parts of the country, and the 1990 Persian Gulf crisis compounded the already volatile situation. The crisis caused oil prices to rise, substantially increasing the cost of oil imports, and foreign exchange earnings to drop. India's creditworthiness, already under strain, became even more vulnerable as Indians from abroad withdrew their substantial foreign currency deposits and commercial banks reduced their exposure. Toward the end of 1990, India's credit worthiness was downgraded, effectively cutting its access to sources of commercial credit. By early 1991, India was on the brink of default.

As the crisis unfolded the debates in India's political and economic circles increasingly focused on reforms. In India's large and highly diverse democracy, those debates proved important in building political consensus around the voices for reforms. Nevertheless, it took a new government, which came to power in June 1991, to launch India's first comprehensive economic policy reform program, which the World Bank supported with a $500 million Structural Adjustment Operation (SAL), approved in December 1991 and closed in December 1993.

Features of Existing Agricultural Policies

The existing policy framework for agriculture is the outcome of many years of experimentation. The evolution of policy and current policy framework can easily be discerned from the changes in objectives. The strategic objectives of agricultural development in India and changes there on can be identified as follows:

Table 1.2 Objective of Agriculture Development Policy in India

Period	Strategic Objectives
Before independence up tomid-60s	To keep prices of food grains low.
Mid 60s to early 80s	Maximizing production of food grains
Early 80s to early 90s	Evolving production pattern according to varying demand patterns.
	Slow opening up of trade in agricultural commodities.

Source: Compiled by the author

Several policy instruments for achieving the above set of objectives have been tried and used in India. The instruments, which are currently in use, include the following:

1. Fixation and announcement of Minimum Support Prices for 24 commodities before sowing and maturing, arrangements for purchases of farm produce at these prices in case market price dip below these levels.
2. Selective intervention in the market for some commodities under the market intervention scheme of the GoI.
3. Open market operation by public agencies and cooperatives for some commodities like raw cotton, oil seeds and copra.
4. Buffer stocking of food grains specially wheat and rice.
5. Public distribution of certain commodities like wheat and rice at subsidized prices.
6. Levy of rice mills and sugar factories and distribution of levy rice and sugar at subsidized prices.
7. Imposition of stock limits on traders and processors.
8. Regulation of marketing practices in agriculture produce markets.
9. Preserving quality and grade standards of agriculture produce markets.
10. Creation of infrastructure facilities for improving marketing such as market yards and sub-yards in primary produce markets roads, communication facilities and dissemination of market infrastructure.
11. Encouraging cooperatives in agricultural development and marketing.
12. Regulation of exports and imports and Agricultural Export Policy

India's Agreement on Agriculture at WTO

After over 7 years of negotiations the Uruguay Round multilateral trade negotiations were concluded on December 15, 1993, and were formally ratified in April 1994 at Marrakesh, Morocco. The WTO Agreement on Agriculture was one of the many agreements which were negotiated during the Uruguay Round.

The WTO Agreement on Agriculture contains provisions in 3 broad areas of agriculture and trade policy: market access, domestic support and export subsidies.

The implementation of the Agreement on Agriculture started with effect from 1.1.1995. As per the provisions of the Agreement, the developed countries would complete their reduction commitments within 6 years, i.e., by the year 2000, whereas the commitments of the developing countries would be completed within 10 years, i.e., by the year 2004. The least developed countries are not required to make any reductions. The products which are included within the purview of this agreement are what are normally considered as part of agriculture except that it excludes fishery and forestry products as well as rubber, jute, sisal, abaca and coir.

Farmers' Suicides in India

Suggestions for the Prevention of Distress in the Farming Community

On account of continuing distress, caused largely by the failure of markets and consequent increase in indebtedness, farmers in the country have been taking recourse of the extreme step of ending their lives. Cases of suicides have been reported from states such as composite Andhra Pradesh, Karnataka, Maharashtra, Kerala, Punjab, Rajasthan, Odisha and Madhya Pradesh. A report of the National Commission on Farmers (NCF, 2006) has underlined the need to address the farmer suicide problem on a priority basis. Some of the measures suggested include:

1. Farmers to be given orientation and awareness to overcoming their stress.
2. Measures by the government to reduce costs of cultivation process and it also can purchase the agricultural produce at a higher price.
3. Import of new technological tools to increase the shelf-life of perishable products.
4. Joining all the rivers of India.
5. Enactment and implementation of strict environment protection laws to be implemented by the government.
6. Provision by the government of affordable health insurance and revitalization of primary healthcare centres. Extension of the National Rural Health Mission to suicide hotspot locations on a priority basis.
7. Setting up State level Farmers' Commissions with representation of farmers for ensuring dynamic government response to farmers' problems.
8. Restructuring microfinance policies to serve as Livelihood Finance, i.e., credit coupled with support services in the areas of technology, management and markets.

9. Cover all crops by crop insurance with the village and not merely the block as the unit for assessment.

10. Provision of a social security net with provision for old age support and health insurance.

11. Promotion of aquifer recharge and rainwater conservation. Decentralization of water use planning. Every village should aim at Jal Swaraj with Gram Sabhas serving as Pani Panchayats.

12. Ensuring availability of quality seed and other inputs at affordable costs and at the right time and place.

13. Recommending low risk and low-cost technologies which can help to provide maximum income to farmers because they cannot cope with the shock of crop failure, particularly those associated with high cost technologies like Bt cotton.

14. Introduction of focused Market Intervention Schemes (MIS) in the case of life-saving crops such as cumin in arid areas. Establishing a Price Stabilization Fund in place to protect the farmers from price fluctuations.

15. Swift action on import duties to protect farmers from fluctuations in the international markets.

16. Setting up Village Knowledge Centres (VKCs) or Gyan Chaupals in the farmers' distress hotspots to provide dynamic and demand driven information on all aspects of agricultural and non-farm livelihoods and also to serve as guidance centres.

17. Conduct of public awareness campaigns to make people identify early signs of suicidal behaviour.

18. Promotion of policies of Integrated Pest Management (IPM) to prevent pest damage – an all-inclusive approach that integrates biological, chemical, mechanical and physical methodology should be used to prevent crop damage.

19. Lowering of fertilizer costs, helping fertilizer industries cut down on costs.

20. Leveraging advancements in Science and Technology by ensuring that state seed policies focus on new genotypes, Contract farming and sensitization to adverse weather conditions.

21. Introduction of the Fasal Bhima Yojana, a crop insurance scheme Paramparagat Krishi Vikas Yojana, to encourage traditional and sustainable organic farming.

Sources of Farmers' Income Growth

There are several possibilities for increasing the income of farmers. Some may be low hanging fruit, i.e., they can be realised within a year or so. Others may be realised over the years depending on the process involved. Also, the strategies and the blueprints should be different for different states/regions and the clientele groups. In fact, one should go to the district or agro-climatic zone level data to understand and suggest location-specific strategies. Abroad scheme for enhancing farmers' income is given below.

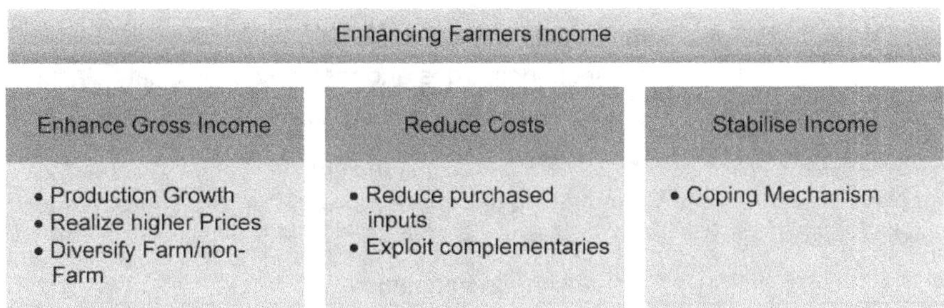

Figure 1.2 Enhancing Farmers' Income

Fundamentally there are three options available for enhancing the income of farmers, viz., increasing the gross income, reducing the costs and stabilising the income. Various options under each of these three broad options are listed.

Table 1.3 Possible options for enhancing farmers' income

Enhancing Gross Income	
Increasing production	• Area expansion – cropping intensity • Yield enhancement – bridging existing yield gaps, enhancing genetic yield potential through breeding/biotechnology, providing irrigation, additional nutrient use
Enhancing output price realization	• Increase in Minimum Support Price (MSP) • Increase in procurement at MSP through market support operations
	• Aggregation for building volume for better price realisation through collective bargaining • Staggered sales and discouraging distress sales and improving holding capacity of farmers
Diversifying within the farm sector	• Choose an ideal farming system considering linkages to increase and stabilise income • Optimal crop mix including horticulture, sericulture, fisheries, animal farming, poultry • Increase scale, increase productivity, enhancing prices of outputs
Diversification to non-farm sector	• Non-Farm Sector (NFS) linked to agriculture – artisans, equipment repairers, tool/implement manufacturers, traders, processing units, custom hiring services • NFS not linked to agriculture – manufacture of consumer goods, shops, hotels, handlooms, etc. • Increase wage income

Table 1.3 *Contd...*

Enhancing Gross Income		
Reducing costs	•	Reducing use of purchased inputs
	•	Input saving technology better agronomical practices
	•	Organic Farming
	•	Better nutrient management through soil health monitoring
	•	Use of bio-pesticides and restoring ecological balance
	•	Farming systems approach/ better by product management
Stabilising income and risk mitigation	•	Traditional coping mechanism
	•	Water saving technology to expand irrigation cover
	•	Crop and asset insurance
	•	Climate change adaptation

Strategies for Doubling Farmers Income

1. Laying emphasis on irrigation along with end to end solution on creation of resources for 'More crop per drop' Krishi SinchayiYojana.
2. Provision of quality seeds and nutrients according to the soil quality of each farm.
3. Promoting large investments in warehouses and cold chains to prevent Post-harvest losses.
4. Promotion of value addition through food processing.
5. Implementation of National Agricultural Markets and e-platforms (e-NAM).
6. To mitigate the risk, introduction of crop insurance scheme at a lower cost.
7. Promotion of allied activities such as Dairy-Animal Husbandry, Poultry, Beekeeping, Horticulture, and Fisheries.

Doubling Farmers Income, Operation Greens and other Recent Policy Responses

Soon after independence, in addition to prioritising food security and low prices, the GoI recognised the need for generating adequate remuneration for farmers. One key component of the Green Revolution interventions was the Minimum Support Price (MSP) for several commodities, which is updated twice a year. The central government has announced that it will set MSPs at 150% of the cost of inputs in line with the aim of doubling farmers' incomes by 2022.

1.2   Revolutions in Agriculture

1.2.1   The Green Revolution

Before focusing on the agricultural sector development in India, let us first look at briefly the overall economic development process of the country since independence in 1947 until the present day. Figure 1.3 illustrates the economic growth rates (three-year moving averages) of India in order to eliminate year to year fluctuations. It is found from the figure that India registered a relatively low economic growth rates of around 3.5 percent per annum until the late 1970s, with large fluctuations owing to the influence the agricultural sector growth which largely depended on the monsoon situation. Indian economy then experienced some improvement in the 1980s because of the government's liberalization policies (but not on a full-scale) under the Rajiv Gandhi regime and a relatively high growth rate attained by the agricultural sector in the decade. And finally, after the full-scale global economic liberalization in 1991 the economic growth rates in India accelerated to a very high level (usually more than 6 percent, and even more than 8 percent after the mid-2000s) until recently.

It is notable at the same time that the agricultural sector growth started to clearly lag behind the GDP growth since the 1990s, which indicates that the Indian economy was plunged into a new developmental stage after the 1990s, where widening disparity between agricultural and non-agricultural (or between rural and urban) sectors has become one of the major problems for the economy. (*Now let us look into the agricultural sector development in India by dividing the whole period from the independence to the present time into several periods.*)

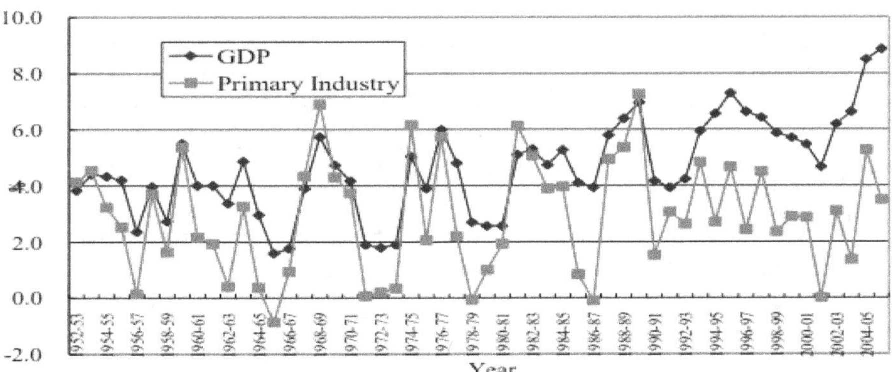

Figure 1.3 Economic Growth Rates in India (Three Year Moving Averages)

Source: Koichi Fujita, 2009, Green Revolution in India

Trends in Food Grain Prices

Trends in Food grain Prices Based on Wholesale Price Index (WPI) (2011-12=100), WPI in the case of food grains increased by 10.38 percent in July, 2019 over July, 2018. Among food grains, WPI of pulses, cereals and oilseeds increased by 20.08 percent, 8.60 percent, and 8.70 percent, respectively, in July, 2019 over July, 2018. Among cereals, WPI for wheat and paddy increased by 5.79 percent and 3.16 percent, respectively, in July, 2019 over July, 2018. Similarly, WPI in case of food grains increased by 1.29 percent in July, 2019 over June, 2019.

Among food grains, WPI of pulses, cereals and oilseeds increased by 0.07 percent, 1.52 percent and 0.001 percent, respectively, in July, 2019 over June, 2019. Among cereals, WPI for paddy and wheat increased by 0.95 percent and 1.72 percent, respectively, in July, 2019 over June, 2019.

Factors that are extended to trigger the growth of the agriculture sector in India

Sustainable growth in the agriculture sector in India will substantially depend on many factors that include:

Increase in demand for food items- With the rising population, (India's projected population in 2030 would reach 1.5 billion) there has been a steady increase in the demand for food items such as milk, vegetables, fruits, rice, wheat and pulses, among others. The projected demand for food grains is mentioned in the table below:

Table 1.4 Food item wise increase in the demand

Types of food items	Demand in year 2000 (in million tones)	Projected demand in year 2030 (in million tones)
Milk	76	182
Vegetables	93	180
Fruits	43	110
Food grains	192	335
Rice	81	156
Wheat	64	95
Cereals	33	102
Pulses	14	30

Source: Industry reports

Food Supply Projections and Gap

The supply for different commodities has been projected using the Triennium Ending (TE) 2010 as the base year production. The supply projections for different food commodities under different scenarios have been presented at 10-year intervals from 2010-2030. To provide a glimpse, food supply and demand gaps for food grains, edible oils and sugar are presented and for high-value commodities, viz. vegetables, fruits, milk, meat, eggs and fish, are given as follows:

Investment in Irrigation Infrastructure

The largest increase in the irrigated area across the world in the next few years is expected in India, with 17.3 M Ha, as public investment in irrigation has remained relatively strong and private investment in groundwater has been rapid. However, even in India, the projected 1995 to 2020 rate of growth in the irrigated area of 1.2 % per year is well below the rate of 2.0 % per year during 1982-93. As we will see in India, the per annum increase in canal irrigated areas is rapidly falling.   The most severe problem facing Indian canal irrigation is not so much the slowdown in its growth, but the rapid deterioration of systems that have already been created. Maintenance is being woefully neglected, leading to poor capacity utilisation, rising incidence of water logging and salinity and lower water use efficiency (WUE). On the whole large canal based irrigation is threatening to become unsustainable physically, environmentally as well as financially.

Various projections have been made for future water demands. The National Commission for Integrated Water Resources Development Plan appointed by the GoI in 1996 has made projections for future water demands for irrigation.

Table 1.5 Future Water Demand Projections in Billion Cubic Meters (BCM)

Scenario	Year 2010	Year 2025	Year 2050
Low	489	619	830
Medium	536	688	1008
High	556	734	1191

Source: Government of India, 1999b:8-9

Note:

1. Low, Medium and High scenarios represent Low, Medium and Medium population projects and 4, 4.5 and 5% growth rates in expenditure.
2. Figures above should be compared with the total average utilisable water resources of 1086 BCM per annum. In 1990, availability level was 520 BCM per annum.

Increased Investment in Agriculture and Infrastructure

We have noted that earlier that public investment in agriculture has been declining, resulting in declining productivity and low capital formation in the agriculture sector. With the burden on productivity - driven growth in the future, this worrisome trend needs to be reversed. Private investment in agriculture has also been slow and must be stimulated through appropriate policies. Accelerated investment are needed to facilitate agricultural and rural development through:

- Introducing new varieties of crops, breeds of livestock, strains of microbes and efficient packages of technologies, particularly those for land and water management, can increase productivity, apart from addressing biotic, socio-economic and environmental challenges;

- Efficient post-harvest and value-addition technologies will obviously help.

- Ensuing the reliable and timely availability of quality inputs at reasonable prices, functioning in place institutional and credit organising support, especially for small and resource-poor farmers, and organising support to land and water resources development are other measures that can be adopted to increased yield.

- Creation of increased employment opportunities, in the rural areas including through creating agriculture-based rural agro-processing and agro-industries, improved rural infrastructures, including access to information, and effective markets, farm to market roads and related infrastructure;

- Centre-staging to the needs and ensuring participation of women farmers is another important area.

Agriculture Exports

India has played an integral role in global agricultural trade in the past one decade. As per the World Trade Organisation data for 2015, it ranks ninth among the world's major exporters. The country produces nearly 600 million tonnes (mt) of agri-products, including horticultural produce. Total agricultural exports registered a CAGR of 16.45% over FY10-18 to reach US$ 38.21 billion in FY18; it stood at US$ 15.67 billion in April-August 2018.

Globally, India is the second-largest producer of rice, wheat, and other cereals. Cereals export stood at US$ 8 billion during 2017-18, with rice (including basmati and non-basmati varieties) contributing a major share. During the same period, fruits and vegetables worth US$ 1.4 billion and spices worth US$ 3.1 billion were exported. Tea exports reached a 36-year high of 240.68 million kg in CY 2017, while coffee exports touched a record 395,000 tonnes. vis-à-vis organic products, India enjoys a twin advantage-an agro-climatic zone suiting varied crop varieties and a rich and native tradition of organic farming-which it is leveraging to meet the growing demand in the domestic as well as international markets.

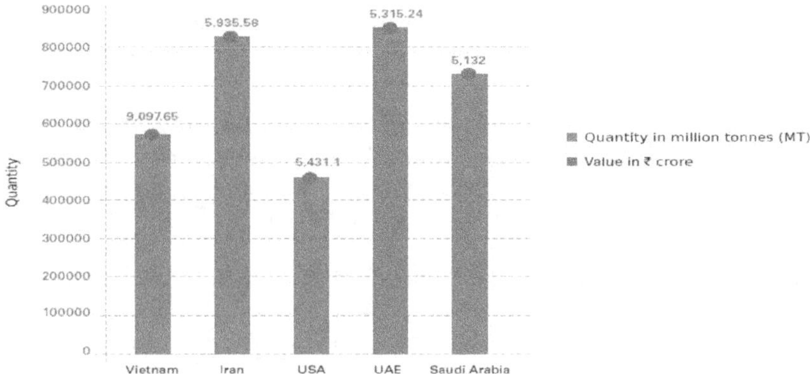

Figure 1.4 Top Five Countries for India's Agricultural Exports, 2018-19

*Source:*http://agriexchange.apeda.gov.in/indexp/genReport_combined.aspx#content

Supportive policy measures have added strength to the geographic advantage the country enjoys, which lends it easy connectivity to Europe, the Middle East, Japan, Singapore, Thailand, Malaysia, and Korea.

A number of factors have contributed to India's rise as a consumer market as well as an export hub—a strong consumer base, evolving food choices, large expanses of farm lands, and a large population engaged in agricultural work. The Make in India drive has already brought in a renewed focus on value addition and on processed agricultural products. The agri-export policy goes several steps ahead, adopting a more holistic approach with implications across the spectrum—production, processing, transportation, infrastructure, crossing the hurdles of competition in quality pricing, R and D, removal of trade barriers, and enabling market access aimed at integration with the global value chain.

1.2.2 The Yellow Revolution – Oil Seeds

In India, oilseeds production, which was only 5.16 million tonnes (mt) in 1950-51, increased to 10.60 mt in 1975–1976. Total oilseeds production increased from 9.37 mt in 1980–1981 to 12.95 mt in 1984–1985. The production, however, again decreased from 12.95 to 10.83 mt in 1985–1986. The spectacular success of the "Yellow Revolution" was observed when oilseeds production rose to 18.03 million tonnes (mt) in 1988–1989 and 24.38 million tonnes (mt) in 1996–1997 and 24.75 million tonnes (mt) in 1998–1999. This spectacular increase in oilseeds production could be attributed to the advent of the "Technological Mission on Oilseeds" in May 1986, which aimed at accelerating self-reliance in oilseeds production in the country. Total Oilseeds production in the country during 2019-20 is estimated at 34.19 million tonnes which is higher by 2.67 million tonnes than the production of 31.52 million tonnes during 2018-19. Further, the production of oilseeds during 2019-20 is higher by 4.54 million tonnes than the average oilseeds production.

Table 1.6 Demand for Edible Oil and Oil Seeds based on Behavioural Approach (Household and Total Consumption) (Million tonnes)

Year	Household Consumption		Total Consumption	
	Edible Oil	Oilseeds (28%)	Edible Oil	Oilseeds (28%)
Baseline (growth rate of GDP at 6% per annum)				
2011-12 Actual	10.66	38.10	9.55	34.11
2012-13	13.27	47.40	11.52	41.13
2016-17	13.27	47.40	14.75	52.67
2020-21	15.82	56.50	17.79	63.54
2021-22	16.57	59.20	18.69	66.76
2028-29	22.73	81.20	26.22	93.65
2032-33	27.13	96.90	31.78	113.49
2033-34	28.36	101.30	33.36	119.13
High growth (growth rate of GDP at 8% per annum)				
2012-13	11.02	39.40	11.52	41.13
2016-17	14.18	50.70	15.67	55.95
2020-21	17.85	63.80	19.83	70.80
2021-22	18.96	67.70	21.08	75.30
2028-29	28.66	102.40	32.16	114.85
2032-33	36.21	129.30	40.86	145.93
2033-34	38.40	137.20	43.40	154.99

Source: Demand and Supply Projections towards 2033, NITI Aayog Working Group Report

Oilseed crops are the second most important determinant of an agricultural economy, next only to cereals within the segment of field crops. The self-sufficiency in oilseeds attained through the "Yellow Revolution" during the early 1990s, could not be sustained beyond a short period. Despite being the fifth largest oilseed crop producing country in the world, presently India is also one of the largest importers of vegetable oils. There is a spurt in vegetable oil consumption in recent years in respect of both edible as well as industrial usages. The demand-supply gap in the edible oils has necessitated huge imports for accounting for 60 per cent of the country's requirement (2017-18: import 15.36 million tonnes; cost Rs. 74,995 crores). Despite the commendable performance of domestic oilseeds production of the nine annual crops (Compound Annual Growth Rate of 3.89%), it could not match with the galloping rate of per capita demand on account of enhanced per capita consumption driven by an increase in population and enhanced per capita income.

As estimated by NITI Aayog Working Group with the increase in population, demand for edible oils is expected to increase. Positive elasticity indicates that growth in income would work as an accelerator for future demand not only for edible oils but also for the other high value commodities. The edible oils demand for household consumption is expected to increase from 11 million tonnes in the base year (2011-12) to 15 million tonnes by the end of 2020-21 and up to 27 million tonnes by 2032-33 with the baseline growth rate of 6 percent per annum. However, if the future economy grows at a higher growth rate of 8 percent annum, the demand for edible oils for household consumption would be up to 18 million tonnes by the end of 2020-21 and 36 million tonnes annum in the baseline scenario and 5 percent in the high growth scenario. Thus, the demand for food grains for household consumption.

1.2.3 The White Revolution

The White Revolution in India occurred in 1970 when the National Dairy Development Board (NDDB) was established to organize dairy development through co-operative societies. Dr. Verghese Kurien was the father of the White Revolution in India. The dairy development program through co-operative societies was first established in the state of Gujarat. The cooperative societies were most successful in the Anand District of Gujarat. The co-operative societies are owned and managed by the milk producers. These co-operatives apart from financial help, also provide consultancy. The increase in milk production has also been termed as Operation Flood.

Operation Flood, launched in 1970 is a project of the National Dairy Development Board (NDDB), which was the world's biggest dairy development program, that made India, from a milk-deficient nation to the world's largest milk producer, surpassing the USA in 1998, with about 17 percent of global output in 2010–11, which made dairy farming India's largest self-sustainable rural employment generator. It was launched to help farmers direct their own development, placing control of the resources they create in their own hands.

The Anand pattern experiment at Amul, a single, cooperative dairy, was the engine behind the success of the program. Verghese Kurien was made the chairman of NDDB by the then Prime Minister of India, Shri Lal Bahadur Shastri, and he was the chairman and founder of Amul as well. Kurien gave the necessary thrust using his professional management skills to the program.

Dairying and Animal Husbandry

Dairying as an economic activity is significant particularly from the viewpoint of rural women for whom it provides them an opportunity for economic empowerment. Many cooperatives have established cattle feed manufacturing plants to meet the demands of the dairy owners for well formulated feed nutrients and to provide information on the health of the animals. The Directorates of Animal Husbandry in the states also have the mandate of providing information and health care services. However, there are variations in milk production as well as per capita in India.

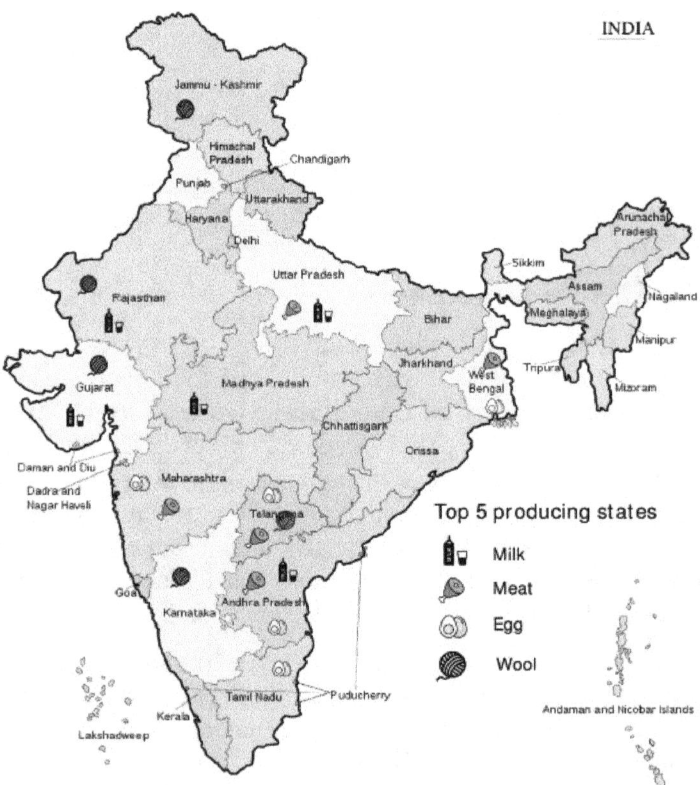

Figure 1.5 Top five milk, meat, egg, and wool producing states in India

In the recent past the dairy sector has transformed into a major activity from its subsidiary status as the value of milk output has surpassed that of output of cereals and pulses combined.

Meat remains the largest export earner among various agriculture commodities. As a result, it is important to understand the value chain of various animal husbandry activities and identify financing opportunities for the banking community in dairy as well as sheep, goat and piggery farming. It is also very important to make strategic plans for dovetailing the macro level infrastructure and marketing plans with the micro-level production activities.

Milk and Milk products are essential food items and provide sufficient essential nutritional support to children. In the context of the GoI's objective of "Doubling of Farmers' Income by 2022" with diversification as one of the key strategies for achieving the same, the animal husbandry sector plays a very important role. This sector provides multiple employment opportunities to small and marginal farmers and plays a major role in ensuring the food security of the most vulnerable part of our population particularly from the point of view of climate change. Therefore, future animal husbandry interventions should focus on climate smart livestock farming that can meet the challenges of climate change and its impacts. There is also a need for understanding the latest trends in livestock farming, pooling of the knowledge sources and formulating strategic approaches for the development of the sector.

Animal husbandry and dairying activities, along with agriculture, continue to be an integral part of human life since the process of civilization started. These activities have contributed not only to the food basket and draught animal power but also to maintaining ecological balance. They play a significant role in generating gainful employment in the rural sector, particularly among the landless, small and marginal farmers and women, besides being sources of cheap and nutritious food.

Increase in Demand for Food Items

Demand for horticultural products including fruits and vegetables will increase from 128.25 million tonnes in the base year of 2011-12 to around 190.5 million tonnes by 2020-21 and further to 462.45 million tonnes by 2033-34 in the baseline scenario. In the high growth scenario, demand will increase up to 343.14 million tonnes in 2020-21 and cross 690.60 million tonnes by 2032-33. The growth in demand for these high value commodities for household consumption will be slightly less than 5 percent per annum in the baseline scenario that would exceed 6 percent per annum in the high growth scenario. The aggregate demand for milk and milk products can touch 355 million tonnes in 2032-33, while meat and meat products can rise up to 44.74 million tonnes.

The aggregate demand for fruits and vegetables could be as high as 580 to 690 million tonnes depending on various growth scenarios.

Table 1.7 Demand for dairy and horticulture based on Behavioural Approach – Household Consumption (million tonnes)

Year	Milk and Products	Egg, Fish and Meat	Vegetables	Fruits	Nuts	Fruits and Vegetables
Actual Demand						
2011-12	69.17	10.11	102.2	26.05	1.64	128.25
Baseline (growth rate of GDP at 6% per annum)						
2012-13	71.48	10.45	105.67	26.94	1.7	132.61
2016-17	85.66	12.52	127.33	32.41	2.04	159.74
2020-21	101.66	14.86	151.89	38.61	2.43	190.5
2021-22	106.39	15.56	159.19	40.44	2.55	199.63
2028-29	144.49	21.13	218.46	55.28	3.48	273.74
2032-33	171.43	25.07	260.74	65.82	4.14	326.56
2033-34	178.96	26.17	272.6	68.77	4.33	341.37
High growth (growth rate of GDP at 8% per annum)						
2012-13	71.48	10.45	105.67	26.94	1.70	132.61
2016-17	91.35	13.36	136.37	34.58	2.18	170.95
2020-21	114.19	16.70	171.91	43.41	2.73	215.32
2021-22	121.07	17.71	182.69	46.07	2.90	228.76
2028-29	180.28	26.38	276.52	69.13	4.34	345.65
2032-33	225.61	33.02	349.29	86.89	5.45	436.18
2033-34	238.69	34.93	370.42	92.03	5.77	462.45

Source: Demand and Supply Projections towards 2033, NITI Aayog Working Group Report

Table 1.8 Demand for dairy and horticulture based on Behavioural Approach – Total Consumption (million tonnes)

Year	Milk and Products	Egg, Fish and Meat	Vegetables	Fruits	Nuts	Fruits and Vegetables
Actual Demand						
2011 -12	119.38	11.64	152.14	88.13	1.64	240.27
Baseline (growth rate of GDP at 6% per annum)						
2012 -13	123.46	12.08	156.21	80.9	1.7	237.11
2016 -17	149.47	14.68	185.39	102.24	2.04	287.63
2020 -21	177.82	18.15	220.41	122.73	2.43	343.14
2021 -22	185.99	19.21	230.65	128.26	2.55	358.91
2028 -29	252.96	28.8	314.33	173.97	3.48	488.3
2032 -33	300.89	36.8	374.15	206.84	4.14	580.99
High growth (growth rate of GDP at 8% per annum)						
2012 -13	123.46	12.08	156.21	80.90	1.70	237.11
2016 -17	155.16	15.51	194.43	104.41	2.18	298.84
2020 -21	190.35	19.99	240.44	127.53	2.73	367.97
2021 -22	200.67	21.37	254.16	133.89	2.90	388.05
2028 -29	288.74	34.05	372.39	187.82	4.34	560.21
2032 -33	355.06	44.74	462.70	227.90	5.45	690.60

Source: Demand and Supply Projections towards 2033, NITI Aayog Working Group Report

1.2.4 The Blue Revolution

The Blue Revolution was launched in India during the 7th Five Year Plan (FYP) that went from 1985 to 1990, during when the government started the Fish Farmers Development Agencies (FFDA)s. During the 8th FYP, from 1992-97, the Intensive Marine Fisheries Program was launched in which collaboration with MNCs was encouraged. Over a period of time, fishing harbours in Tuticorin, Porbandar, Visakhapatnam, Kochi and Port Blair were established. A number of research centres have also been set up to increase the production as well as to do improvement in species.

The Blue Revolution is part of the Government's efforts to promote fishing as an allied activity for farmers in order to double their incomes. It refers to an explosive growth in the aquaculture industry. As part of its efforts to raise seafood output and exports and promote sustainable aquaculture, the Government has constituted an independent Ministry for Fisheries. In the budget 2019-20, the government allocated an estimated 3,737 crore rupees for the newly carved out Ministry of Fisheries, Animal Husbandry and Dairying.

Fish touch our lives in countless ways in terms of providing food, nutrition, recreation, livelihood, employment and many more. It comes mainly from two modes of production systems: Capture Fisheries (capturing wild fish from marine and freshwater) and Culture Fisheries (farming fish, also known as aquaculture). The total fish production of 12.59 million metric tonnes was registered during 2017-18 with a contribution of 8.90 million metric tonnes from the inland sector and 3.69 million metric tonnes from the marine sector. The average growth in fish production during 2017-18 stands at 10.14 per cent when compared to 2016-17 (11.43 million metric tonnes). This is mainly due to 14.05 per cent growth in Inland fisheries when compared to 2016-17 (7.80 million metric tonnes). India is currently the world's second-largest producer of fish. It is also world number two in aquaculture production as well as in inland capture fisheries. India is the second-largest fish producing country in the world with an annual production of about 12.60 million metric tonnes; it is aimed to increase it to 15.00 million metric tonnes by 2020 of which 65 per cent was from the inland sector. Almost 50 per cent of inland fish production is from culture fisheries, which constitutes 6.5 per cent of global fish production. The sector has been showing steady growth in the total gross value added and accounts for 5.23 per cent share of agricultural GDP. Fish and fish product exports emerged as the largest group in agricultural exports and in value terms accounted for Rs. 47,620 crore in 2018-19. As per The Economic Survey 2018-19, Foreseeing the vast resource potential and possibilities in the fisheries sector, a separate Department of Fisheries was created in February 2019. The Government has merged all the schemes of fisheries Sector into an umbrella scheme of 'Blue Revolution: Integrated Development and Management of Fisheries focusing on increasing fish production and productivity from aquaculture and fisheries resources, both inland and marine".

The growth in fish production in India has been at a faster rate than in the world in general; mainly as a result of the increasing contribution from inland fisheries. In the pre-WTO period of 1990-91, the share of exports India's fisheries was 3.90, which rose to 4.56 in the post-WTO period of 2003-04.

Overall, the share of the developing world in total world fish production increased from 43 percent in 1973 to about 73 percent in 1997.Mainly on account of increased contributions from countries such as China and India.

From a production level of 0.75 million tonnes during 1950-51, fish production reached 13.34 million tonnes during 2018-2019. With a share of 6.30% in global fish production and 5% in global trade, India is the 2nd largest fish producer in the world in terms of total production and also ranked 2nd in aquaculture production. While the growth in marine sector is stagnating with a CAGR of 2.5%, the inland sector has been growing at a CAGR of 5.74% (Fig 13) supported by the growth in aquaculture production especially of carps, pangaisus fish and shrimp.

Trend in Export of Marine Fish

Total fish production in India was 12.59 million tonnes (mt) and the country exported 1.38 mt fish and fish products with a value of over ₹45,000 crore in 2017-18. Inland capture fisheries accounted for 8.9 mt, the share of marine fisheries was 3.69 mt. India is currently the second largest producer of fisheries after China. The total fish production in 2017-18 was 10.14 per cent more than 11.43 mt produced in 2016-17. Inland fisheries, which grew at 14.05 per cent accounted for much of the growth. Marine fisheries production, on the other hand, went up by only 1.73 per cent in 2017-18.

Table 1.9 Top Five fishery States in India

States	Inland (in lakh tonnes)	Marine (in lakh tonnes)	Total (in lakh tonnes)
Andhra Pradesh	28.45	6.05	34.50
West Bengal	15.57	1.85	17.42
Gujarat	1.34	7.01	8.35
Kerala	5.34	1.51	6.85
Tamil Nadu	1.85	4.97	6.82
All India Production	89.02	35.88	125.90

Source: State Governments, CIFRI & CMFRI

Andhra Pradesh, which captured 3.45 MT of fish, topped the Indian states, followed by West Bengal, which accounted for 1.74 MT. While Andhra Pradesh, retained the top position in inland capture fisheries with 2.85 MT, Gujarat with 7.01 lakh MT was number one in marine fisheries. There has been steady growth in the export of fish and fish products from India.

Table 1.10 Trend in Export of Marine Products

Year	Marine Products Export Value (in ₹ Crore)
2011-12	16,597.23
2012-13	18,856.26
2013-14	30,213.26
2014-15	33,441.61
2015-16	30,420.83
2016-17	37,870.90
2017-18	45,106.89
2018-19	46,589.37
2019-20	45,000.53

Source: State Government, CIFRI & CMFRI

During 2018-19, the volume of fish and fish products exported was 1.39 MT and ₹46,589 crore in value and 2017-18, the volume of fish and fish products exported was 1.38 mt and ₹45,107 crore in value. The export of marine fish products registered an annual growth of 21.35 per cent in volume and 19.11 per cent growth in value.

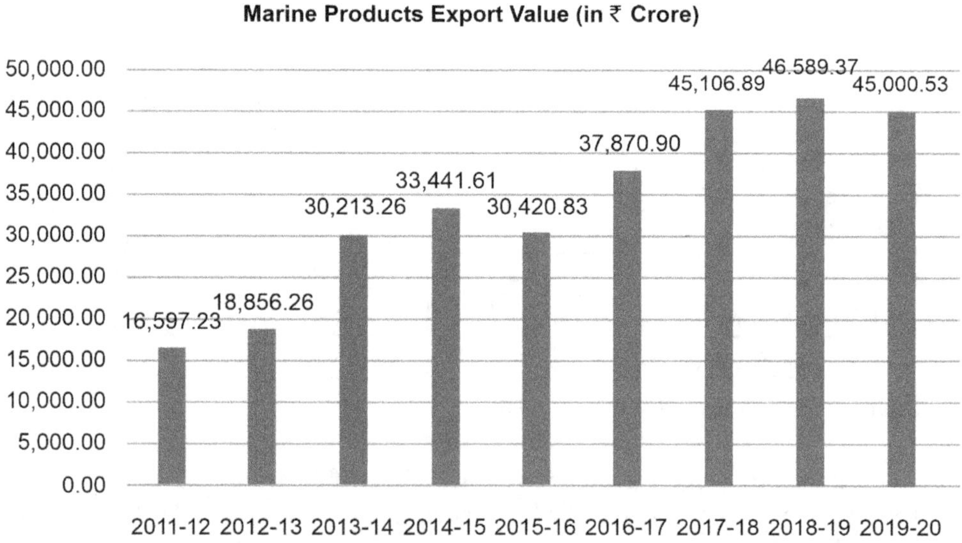

Marine Products Export Value (in ₹ Crore)

Figure 1.6 Marine Products Export Value from 2011 to 2020

Source: State Government, CIFRI & CMFRI

1.3 Current Trends in Agriculture

1.3.1 Government Initiatives

The government has set a goal of doubling farmers' incomes by 2022. The 14-volume report of the Committee of Doubling Farmers Income, headed by Ashok Dalwai, has an entire volume devoted to the need for science and technology interventions in agriculture. The use of Big Data, the Internet of Things, Artificial Intelligence and Block Chain in developing value chains and the use of robots and sensors figure significantly, along with the need for better research in crop science and genetically modified technologies, in the to-do list.

The Indian Council of Agricultural Research (ICAR) 2025 Vision Document, too, emphasises the use of technology, including robotics and automation; energy-efficient and environment-friendly devices for farm operations; GIS, GPS and remote sensing; climate smart resource management technologies; and pathogens. According to the document, high-performance computing could be used to analyse very large data sets, particularly those related to agricultural genomics, proteomics, geo-informatics and climate change. It also articulates the need to have a regulatory process for new technological advancements and suggests that a mix of technologies be used to achieve higher productivity.

Some of the recent major government initiatives are:
- The Pradhan Mantri Kisan Samman Nidhi Yojana (PM-Kisan) under which ₹ 2,021 crores (US$ 284.48 million) was transferred to the bank accounts of more than 10 million beneficiaries on February 24, 2019.
- The Transport and Marketing Assistance (TMA) scheme to provide financial assistance for transport and marketing of agriculture products in order to boost agriculture exports.
- The Agriculture Export Policy, 2018 which aims to increase India's agricultural exports to US$ 60 billion by 2022 and US$ 100 billion in the next few years with a stable trade policy regime.
- The ₹ 15,053 crore (US$ 2.25 billion) procurement policy named 'Pradhan Mantri Annadata Aay Sanraks Han Abhiyan' (PM-AASHA), announced in 2018, under which states can decide the compensation scheme and can also partner with private agencies to ensure fair prices for farmers in the country.
- The ₹ 5,500 crores (US$ 820.41 million) assistance package for the sugar industry in India was announced in 2018. The Pradhan Mantri Krishi Sinchai Yojana (PMKSY) with an investment of ₹ 50,000 crores (US$ 7.7 billion) aimed at the development of irrigation sources for providing a permanent solution from drought.

- Also, on the anvil is a provision of ₹ 2,000 crores (US$ 306.29 million) for computerisation of Primary Agricultural Credit Societies (PACSs) to ensure that cooperatives are benefitted through digital technology.
- A new AGRI-UDAAN programme to mentor start-ups to enable them to connect with potential investors in order to enable them to connect with potential investors in order to triple the capacity of the food processing sector in India from the current 10 per cent of agriculture produce. A commitment of ₹ 6,000 crores (US$ 936.38 billion) as an investment for mega food parks in the country, as a part of the Scheme for Agro-Marine Processing and Development of Agro-Processing Clusters (SAMPADA) allowing.
- Permitting 100 per cent FDI in the marketing of food products and in food product e-commerce under the automatic route.

1.3.2 Production and Trade of Major Farm Products

India's food grain production is expected to reach a record 291.95 million tonnes, as per the second advance estimates of production of food grains, oilseeds and other commercial crops for the agricultural year 2019-2020.As per the second advance estimates for 2019-20, total foodgrain production in the country is estimated at a record 291.95 million tonnes which is higher by 6.74 million tonnes than the production of foodgrains of 285.21 million tonnes achieved during 2018-19. Foodgrain production during 2019-20 is also higher by 26.20 million tonnes compared with the average production during the previous five years (2013-14 to 2017-18).The total production of rice during 2019-20 is estimated at a record 117.47 million tonnes, which is higher by 9.67 million tonnes than the five years' average production of 107.80 million tonnes.

Production of wheat during 2019-20 is estimated at a record 106.21 million tonnes, which is higher by 2.61 million tonnes compared with wheat production during 2018-19 and is 11.60 million tonnes higher than the five-year average wheat production of 94.61 million tonnes. Production of nutri/coarse cereals, estimated at 45.24 million tonnes, is higher by 2.18 million tonnes than the 43.06 million tonnes production achieved during 2018-19. It is also higher by 2.16 million tonnes than the five-year average production. The total production of pulses during 2019-20 is estimated at 23.02 million tonnes, which is higher by 2.76 million tonnes compared to the five-year average production of 20.26 million tonnes. Total oilseeds production in the country during 2019-20 is estimated at 34.19 million tonnes, which is higher by 2.67 million tonnes than the production of 31.52 million tonnes during 2018-19. Further, the production of oilseeds during 2019-20 is higher by 4.54 million tonnes than the average oilseeds production.

Total production of sugarcane in the country during 2019-20 is estimated at 353.85 million tonnes. The production of sugarcane during 2019-20 is higher by 4.07 million tonnes compared to the average sugarcane production of 349.78 million tonnes in the past five years.

Cotton production is estimated at 34.89 million bales (of 170 kg each) in 2019-20, which is higher by 6.85 million bales than the production of 28.04 million bales produced during 2018-19. Production of jute and mesta is estimated at 9.81 million bales (of 180 kg each).

1.3.3 Emphasis on Rain Fed Ecosystems

Resource-poor farmers in the rain fed ecosystems practice less-intensive agriculture, and since their incomes depend on local agriculture, they benefit little from increased food production in irrigated areas. To help them, special effort is needed to disseminate available dry land technologies and to generate new ones. It will be necessary to enlarge the efforts for promoting available dry land technologies, increasing the stock of this knowledge, and removing pro-irrigation biases in public investment and expenditure, as well as credit flows, for technology-based agricultural growth. Farming system research to develop location specific technologies must be intensified in the rainfed areas. Strategy to make grey areas green is expected to lead to a second Green Revolution, which would demand a three-pronged strategy - watershed management, hybrid technology and small farm mechanisation, including land scape farming.

1.3.4 Fertilizer Application to Enhance Soil Health

Building and maintaining soil quality is the basis for harmonious and successful farming. The link between soil quality, farming practices, long-term soil productivity, sustainable land management, agriculture and environmental quality is now widely acknowledged (Fig. 1.7), as it represents the importance of conserving soil as a resource for future generations instead of "soil fertility", which was otherwise associated with crop yield alone.

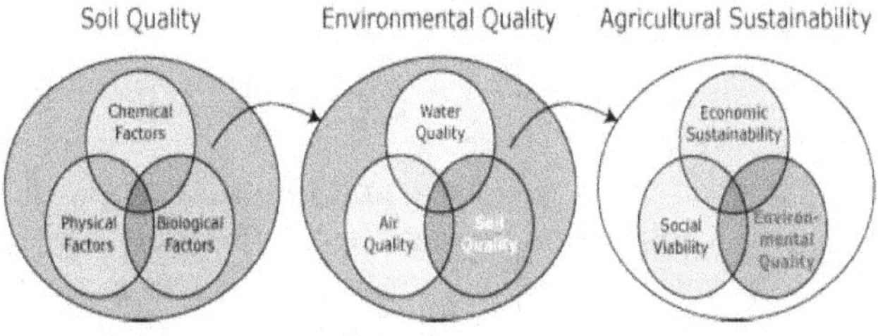

Figure 1.7 Relationship among soil quality, environmental quality and agricultural sustainability

Soil Testing

The main science-based tool to estimate a soil's capacity to supply nutrients on agricultural land is soil testing. As a number of non-fertility factors impact final crop yield (environmental conditions, pests, etc.,) soil testing is not an exact science, and resulting fertilizer recommendations may not be precise. Nevertheless, soil testing remains one of the most valuable and effective nutrient management tools available.

Zero Budget Natural Farming (ZBNF)

ZBNF is a set of practices that aims at optimising farming costs and reducing the gap between promise and actual realisation. Integrates the sustainable farm intensification practices with a focus on minimising the costs. Zero budget means zero cost, i.e., "no need for market based inputs", it is propagated that only 1.5 per cent of the nutrients required by the plants are provided by the soil and the remaining 98.5 per cent comes from air, water and solar energy. Even the 1.5 per cent required is available in plenty in every type of soil, albeit in an unavailable form. Thus, micro-organisms in the soil can be increased with the application of desi (i.e., country like India) cow dung and no fertilizer or pesticide is needed. Zero budget natural farming is based on 4 wheels/ non-negotiable guidelines/principles or package of farming practices that would increase soil health and crop yields at zero external inputs or costs. These include: 1. Jivamrita (life tonic); 2. Bijamrita (seed protection tonic); 3. Acchadana (mulching) and 4. Whapasa (soil aeration/moisture).

1. Jivamrita acts as a catalytic agent that enlivens the soil, increasing microbial activity and organic matter. It helps to prevent fungal and bacterial growth and increases earthworm activity.

2. Bijamrita protects seedlings from seed or soil borne diseases, as well as young roots from fungus.

3. Acchadana enhances decomposition and humus formation through the activity of the soil biota activated by Jivamrita.

4. Whapasa is the condition in which there are both air molecules and water molecules present in the soil.

ZBNF encourages reducing irrigation and recommends irrigating only at noon in alternate furrows. There are also a number of pest management measures such as neemastra, agniastra and brahmastra – which are home-made preparations used for insect and pest control. While the scientific credentials are tested, Andhra Pradesh (AP) is promoting it on a large scale. In fact, AP Government is aiming to cover all the 6 million farmers and 8 million hectares in the state under the initiative of Climate Resilient Zero Budget Natural Farming (CRZBNF) by 2027. In Andhra Pradesh it was observed that yield of various crops have gone up, ranging between 9 per cent (paddy) to 40 per cent (ragi), net incomes have gone up substantially ranging between 25 per cent (ragi) and 135 per cent (groundnut). The main objective of CRZBNF is to make agriculture viable, sustain agrarian livelihoods and reduce agrarian distress through cost reduction and sustainable agricultural practices that are climate resilient. It aims to reduce costs of cultivation and climate risks, enhance yields and soil fertility through the adoption of agro-ecology framework. Extension support is led by farmers (including women) and through farmer-to-farmer learning.

The State Government's target is to reach out to all farmers in the state (6 million including tenants) and achieve 100 per cent chemical free agriculture by 2027. This is an unprecedented transformation towards sustainable agriculture on such a massive scale. Its focus is on the poorest of the poor farmers (bottom 30 per cent - above 1.5 million families) with nutrition and livelihoods security. This transformation is expected to be achieved through providing support for each farmer family for at least 5 years till they attain sustainable and viable livelihoods. CRZBNF also aims to create human and social capital necessary for vibrant and inclusive agricultural production.

Table 1.11 Farmers' perception on CRZBNF vis-à-vis non- CRZBNF

CRZBNF	**Non-CRZBNF**
Farmers see lot of benefits in terms of soil quality and water conservation	Business as usual, soil degradation is a major concern. Increased use of FYM/organic matter
Yields are much lower than expectations Wrong or misinformation about yield rates	Yield fluctuations are often related to climate
Timely availability of inputs. Not very convenient to practice	Easy and convenient due to readymade availability of inputs
Only hardworking farmers with family labour only could practice	Minimum involvement of family labour
Difficult to practice in irrigated areas (canal command) due to sea-page of chemicals	Climate resilient land use practices are limited to dryland areas
Some farmers adopt only for self-consumption, as they don't see much advantage in terms of net economic benefits	Apart from high costs it is more convenient with established set of practices.

Source: Drawn from the discussions with farmers

Climate Resilient Zero Budget Natural Farming (CRZBNF) is the most sustainable practice but the spread is slow. If the climate of increased yields at reduced (if not nil) costs are true, CRZBNF should have spread much faster than Green Revolution technologies where high yields are associated with high costs. It is true that sustainable practices need support and promotion, especially when environmental or/and social benefits outweigh the immediate economic benefit. The promotional strategies need to be designed accordingly rather than based on false expectations. The strategies ought to integrate policies and institutions at central and state levels. The success of Green Revolution technologies lies in such coordinated efforts, viz., input supply chains fostered with input subsidies along with promotional activities. Moreover, the benefit (impact) of adaption are dramatic in the case of Green Revolution technologies setting an effective demonstration effect. This may not be the case with CRZBNF as the process of realising the impact is slow and hence would require direct support in terms of demonstration, awareness building, etc.

System of Rice Intensification (SRI)

The system of rice intensification is an innovative method comprising uncomplicated management practices that allow rice-growers to attain higher productivity. Similar to the central principle of sustainable agriculture which seeks to make optimal use of naturally available resources as functional inputs, SRI too works by integrating invigorating processes such as optimum plant population, transplanting single young seedlings, wider square planting, mechanical weeding at 10 DAT (Days After Transplantation), keeping the soil moist but not inundated, using Leaf Colour Chart (LCC)-based nitrogen management, using compost, Farm Yard Manure (FYM) or green manure to the greatest extent possible and converting all of these resources synergistically to achieve higher yield from the rice crop. The synergy between these practices helps to produce more healthy and productive plant phenotypes and subsequently to obtain higher returns. SRI also minimises the use of high-cost external inputs that are detrimental to farmers and how productively they are able to employ them.

Six Key Elements of SRI Farming

SRI farming practices for traditional rice growing methods are:

1. Transplanting seedlings much earlier than in conventional methods.
2. Planting only one seedling per hill, rather than a handful.
3. Spacing plants wider apart than in conventional methods and arranging them in a square pattern.
4. Applying water intermittently instead of continuous flood irrigation.
5. Using rotary weeding to control weeds and promote soil aeration; and
6. Applying organic fertilisers to enhance soil fertility and yield.
7. The available studies on the SRI method of paddy cultivation suggest that farmers can double paddy productivity with reduced use of farm inputs

and irrigation water. Using the SRI method of paddy cultivation, countries like India, Indonesia, Cambodia, Vietnam, and the Philippines have recorded an increase of rice yield from 60 per cent to over 170 per cent.

Post-harvest Management

India produces, annually, over a billion tonnes of raw food crops and commodities, and some of these, mainly fruits, vegetables, milk, meat and fish are highly perishable. For want of adequate cold chain facilities, and processing and product development technologies, a substantial amount of produce is lost. The country can ill afford this loss. On an average, postharvest losses of 10-25% in durables and 30-35% in fruits and vegetables were documented. The challenge is to handle fresh produce post-harvest with reduced losses, value additions and maintenance of eating quality. Agro-processing is a sunrise sector in the Indian economy, in view of its large potential for growth and likely socioeconomic impact, specifically on employment and income generation. Estimates suggest that, in developed countries, up to 14% of the total workforce is engaged in it directly or indirectly. In India, however, only 3% of the workforce finds employment in this sector, a reflection of its underdeveloped state and vast untapped potential. Skills need to be developed to undertake primary processing or small product development, while the industrial sector needs to look at the large-scale production of value added products with enhanced shelf lives.

1.3.5 Diversification of Agriculture and Value Addition

In the face of shrinking natural resources and the ever-increasing demand for larger food and production due to high population and income growth. Intensification is needed to drive the future growth of agriculture research for product diversification. Besides developing technologies for promoting intensification, greater attention needs to be paid to the development of technologies that will facilitate diversification, particularly towards intensive production of fruits, vegetables, flowers and other high value crops. Per capita availability of arable land is quite low and has been declining over time. Diversification, towards these high value and labour-intensive commodities can provide adequate income and employment to the farmers dependent on the small size of farms. Due importance should be given to quality, nutritional aspects, development of post-harvest handling and agro-processing facilities and value addition technologies. The role of biotechnology in post-harvest management and value addition needs to be enhanced.

Value addition and Cost-Effectiveness

Post-harvest losses generally range from 10 to 25 percent for non-perishables and about 30 to 35 per cent in fruits and vegetables. These losses can and must be minimized. Emphasis should therefore be placed to develop post-harvest handling, agro-processing and value-addition technologies not only to prevent losses, but also to improve quality through proper storage, packaging, handling and transport. With the thrust on globalization and increasing competitiveness, this approach is expected to improve the agricultural export contribution of India. Agro-processing facilities should preferably be located close to the points of production in rural areas, to facilitate the promotion of off-farm employment. Agricultural cooperatives and Gram Panchayats need to be encouraged to play a leading role in this effort, keeping the needs of small farmers in mind.

Increased Investment in Agriculture and Infrastructure

We have noted earlier that public investment in agriculture has been declining, resulting in declining productivity and low capital formation in the agriculture sector. With the burden on productivity-driven growth in the future, this worrisome trend needs to be reversed. Private investment in agriculture has also been slow and must be stimulated through appropriate policies. Accelerated investment are needed to facilitate agricultural and rural development through:

- Introducing new varieties of crops, breeds of livestock (Gir, Kankrej and Ongole, etc.), strains of microbes and efficient packages of technologies, particularly those for land and water management, can increase productivity, apart from addressing biotic, socio-economic and environmental challenges.

- Promoting efficient post-harvest and value-addition technologies.

- Ensuring the reliable and timely availability of quality inputs at reasonable prices, putting in place institutional and credit organising support, especially for small and resource-poor farmers, and organising support to land and water resources development.

- Creation of increased employment opportunities, in the rural areas including through creating agriculture-based rural agro-processing and agro-industries, improved rural infrastructures, including access to information, and effective markets, farm to market roads and related infrastructure.

- Centre staging the needs, and ensuring participation of women farmers is another important area.

1.3.6 The Farm to the Fork approach and Sustainability

Feeding the burgeoning population while simultaneously ensuring sustainable management of natural resources such as land, water and the atmosphere is a major challenge. Production increases so far have been at the cost of depleting or degrading resources, especially soil, water and biodiversity. Doubling farmers' incomes and sustaining food and nutritional security call for an integrated approach to various sectors of agriculture viz., food grain crops, animal husbandry, dairy farming, horticulture (fruits and vegetables) and forestry, livestock and fisheries, process engineering and machines as against our hitherto major tilt towards food crops over others.

Delivering Farm to Fork Sustainability

Agribusiness is a practice of activities, with backward and forward linkages related to research, extension, financial services, meteorological services, production, storage, processing, marketing, trade, and distribution of raw and processed food, feed and fibre. With the structural transformation of the economy, the share of crop production (farming) is decreasing, and that of processing, distribution and trade is increasing. With rising disposable incomes reflected in the augmented purchasing power and growing nuclear families, the demand for processed foods is witnessing a considerablerise. These factors are further inducing the need for sophistication in various segments of agribusiness such as procurement, storage, transportation, distribution, etc. The demand to feed the growing population and tap emerging global market opportunities shall provide additional thrust to the sector.

Good Agricultural Practices (GAP)

GAP are internationally recognized basic environmental and operative conditions necessary for the production of safe and wholesome fruits and vegetables. They ensure economic and social sustainability for on-farm processes and result in safe and quality food and non-food agricultural products. In the era of declining factor - productivity and degradation of the natural resource base of agricultural production system, there is an urgent need to develop GAP. The challenge, especially in a country such as India, is to transfer GAP to the farmers' field. The present training on GAP includes almost every aspect of farm operations for efficient use of farm resources and a better understanding of them will lead to quality outcomes from field operations. In the context of the extant resource and production vulnerabilities including climate change, GAP will help meet those challenges apart from raising farmers' incomes and employment generation, minimizing the risks in farming and enhancing resource use efficiency.

Importance of Agribusiness in India

Following the structural transformation of the Indian economy, the share of agricultural production (farming) in the economy has been decreasing steadily, while that of processing, distribution and trade is increasing. Further, with the increase in backward and forward linkages, the distinction between agriculture and agro-industry is also blurring. Farm production, processing and trade continue to integrate further. The need for a robust supply chain is gaining momentum.

With rising disposable incomes reflected in the augmented purchasing power and a growing number of nuclear families, the demand for processed foods is witnessing a substantial surge. These factors are spurring the need for sophistication in various segments of agribusiness such as procurement, storage, transportation, distribution, etc.

The value chain of agribusiness

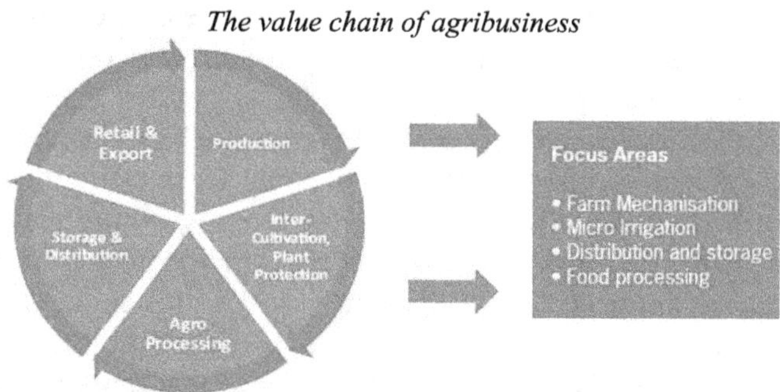

Figure 1.8 Composition of segments in agribusiness in India

Production

Production involves sowing, planting and harvesting of the produce that is to be channelized in the agribusiness value chain. It is crucial to ensure high quality of production as well as higher levels of yield.

Agribusiness Growth and Future Potential

Agribusiness Growth

Over 52.8% of India's arable land can be cultivated, compared to the global average of 11%. India is among the highest-ranking countries in terms of output for various commodities like rice, cotton and dairy goods. On the other hand, there is a need for sustained efforts to improve productivity through measures such as simplification of regulatory processes, enhancement of efficiencies in the food distribution system, spreading awareness on modern agricultural practices and tackling unpredictable weather patterns.

To this end, the GoI has, from time to time launched various initiatives such as establishing mega food parks, initiating easier access to credit for farmers, creating a long-term irrigation fund, introducing Krishi Kalyan Cess and the National Cold Chain Development Scheme, among others. Over the years, India has developed export competitiveness in certain specialised products, making it the world's 15ᵗʰ largest agricultural, fishery, and forestry product exporter.

Small-sized farm holdings are, quite naturally, a challenge when, increased production and enhanced productivity are being addressed. Over 85% of farmers in India belong to the small/marginal category with holdings measuring less than 1 ha. Small and marginal farmers with less than two hectares of land account for 86.2% of all farmers in India, but own just 47.3% of the crop area, according to provisional numbers from the 10th agriculture census 2015-16.Well-intentioned programmes such as the Mahatma Gandhi National Rural Employment Guarantee Act (MNREGA) have also exacerbated the situation by shrinking the availability of labour, apart from resulting in wage levels rising unacceptably. The best way to overcome these challenges would be through the evolution of farm mechanisation.

The agriculture sector in India has witnessed a considerable decline in the use of animal and human power. This has resulted in a shift from traditional agricultural practices to more mechanised processes. Though the level of mechanisation in India is lower as compared to other developed countries, it is certainly showing an upward trend, particularly in farmer's cooperatives.

Table 1.12 The extent of mechanisation at various levels of the value chain

Extent of mechanisation	Levels of value chain
Soil working and seedbed preparation	40%
Seeding and planting	29%
Plant Protection	34%
Irrigation	37%
Harvesting and Threshing	60–70%
Overall	40–45%

Source: Department of Agriculture and Cooperation, Ministry of Agriculture, GoI

The use of farm equipment can increase farm productivity by ~30% and reduce the input cost by about 20%. The government is therefore promoting farm mechanisation by subsidising the purchase of equipment as well as supporting bulk buying through front-end agencies. Considerable emphasis needs to be laid on R and D in these areas and lessons need to be learnt from countries such as Japan.

Micro Irrigation

Given the dependence of Indian agriculture on natural water resources and monsoons, the efficient use of available water resources becomes more crucial. With the use of micro irrigation systems, conveyance losses are minimised.

Evaporation, runoff and deep percolation are also reduced by using micro irrigation methods. Another water saving advantage is that water sources with limited flow rates such as small water wells can also be used. Micro irrigation provides significantly higher water usage efficiency on account of proximity and focused application. Hence, the concept of micro irrigation has been re-introduced under the Pradhan Mantri Krishi Sinchayee Yojana (PMKSY).

Micro irrigation can result in an overall saving of irrigation water by 20-38%, fertilisers by 28.5% and energy by 30.5%. The total potential of micro irrigation in India is estimated at around 69 million hectares, however, the coverage of micro irrigation was only 8.53 million hectares in FY 2019 (drip irrigation coverage being 3.37 million hectares and sprinkler irrigation coverage being 4.36 million hectares). There is thus, a good deal of untapped potential.

Water use efficiency – "Per Drop More Crop"

The GoI has the mandate to increase the water use efficiency by at least 20 per cent. It can be observed that the water use efficiency in terms of biomass produced in kg/m3 of water consumed can be increased by more than 200 per cent in the case of vegetables by using the drip method of irrigation. Similar is the case with fruits, oilseeds, pulses, cash crops such as sugarcane and cotton, etc. If the drip irrigation systems are adopted on a large scale for all the crops, this will not only provide "Har Khet Ko Pani" and "Per Drop More Crop", but will also fulfil our beloved Prime Ministers' dream of "Doubling Farmers' Income" in the near future.

Cold Chains

India is the leading producer for many of the agricultural commodities such as fruits, vegetables, spices, milk and other fishery products. Still, the share of India's exports in these segments is comparatively low. The prime reason for the same is the lack of appropriate cold chain infrastructure facilities which includes both storage and transportation facilities. However, with increasing urbanisation and growth of organised retail, food servicing and food processing sector, there has been a boost in the growth of the cold chain industry.

India's cold chain sector is a combination of surface storage and refrigerated storage. The industry has grown at a CAGR of 20% in the last 3 years (2014-16). The cold chain market in India is expected to reach ₹ 624 billion by the end of 2017. The Indian cold chain market was worth INR 1,121 Billion in 2018. The market is further projected to reach INR 2,618 Billion by 2024, growing at a CAGR of 14.8% during 2019-2024. (The biggest source of revenue for the Indian cold chain industry is the cold stores. There are around 6,300 cold storage facilities in the country, with an installed capacity of 30.11 million tonnes. The current cold storage capacity in India is pegged at 37-39 million tonnes (mt). According to official statistics, there are about 7,645 cold storages in the country with 68 per cent of the capacity being used for potato, while 30 per cent is multi-commodity cold storage. Major players in this market include ColdEx, Bhramanand Himghar, Dev Bhumi Cold Chain, Gati and Snowman Logistics, among others.

Some interesting facts about the cold chain industry in India are (as of 2016):

- Organised players contribute only 8 to 10% of the cold chain industry market
- 36% of cold storages in India have capacities below 1,000 MT and are unevenly spread.
- 65% of India's cold chain storage capacity is contributed by the states of Uttar Pradesh and West Bengal
- At the current capacity less than 11% of what is produced can be stored

Factors that have accelerated the growth of Indian cold chain industry are:

- Growth in the organised retail industry – Over the last few years, organised retail and food service industries have emerged as new segments of the cold chain, mainly due to changing consumption patterns. There is an increased demand not only for capacity addition of cold storage facilities for a set of highly perishable products, but also for a wide variety of vegetables, fruits and grains.
- Growth in end-user segments - With the growth in the end user segment, cold chain infrastructure is expected to get a boost and help in reducing wastage.
- Demand from the pharmaceutical sector – The growth in the pharmaceutical industry has created an incidental demand for the increase in cold chain facilities in the country.

Food Processing

The Food processing industry is known to provide a vital link between the agriculture and manufacturing sectors of the economy.

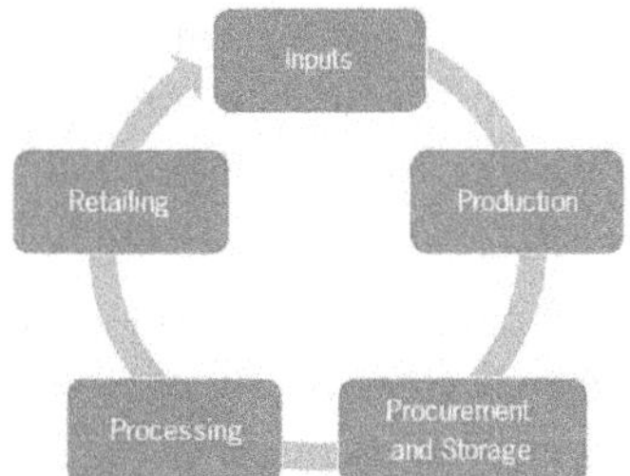

Figure 1.9 The value chain in the food processing sector

There has been a significant increase in the number of registered food processing units in the country i.e. from 26,219 units in 2007-08 to 39,748 units in 2019. The industry engages approximately 1.85 million people in around 39,748 registered units in 2019 with fixed capital of $ 32.75 billion and aggregate output of around $ 158.69 billion. Major industries constituting the food processing industry are grains, sugar, edible oils, beverages and dairy products.

Some of the factors that have contributed to the growth in the food processing sector are:

- Increased FDI inflows – 100% FDI is permitted under the automatic route in food processing industries. As a result of this, the sector has witnessed an increase in FDI inflows which has directly led to the acceleration of the growth of this sector.

- Mega Food Parks Scheme -As a result of the government-led initiative Mega Food Park Scheme, 42 mega food parks are being established in the country with a total investment of ₹ 155 billion. The primary objective of the scheme is to facilitate the establishment of an integrated value chain, with processing at the core, supported by requisite forward and backward linkages. The scheme now covers 22 states of India. Currently, 17 mega food parks have become functional.

- Recognition as a priority sector - The food processing sector was recognised as a priority sector in 2011 to ensure a greater flow of credit to entrepreneurs for setting up food processing units and attracting investment in the sector.

- Creation of the food processing fund - In the budget for the year 2015-16 the GoI allocated a corpus of ₹ 2,000 crore (approximately US$ 300 million) for the creation of a special fund called "Food Processing Fund" under the National Bank for Agriculture and Rural Development (NABARD) to provide cheaper credit to the food processing industry. Excise duty on plant and machinery for packaging and processing was brought down to 6% from 10%.

Policy Initiatives to Support the Growth of Agribusiness in India

There have been a lot of government initiatives and measures to propel the growth of the agriculture sector in the country. The government targets to double the income of farmers by 2022, from the level obtaining in 2016. To achieve this the government has undertaken many initiatives which include:

- Increased importance for micro irrigation – The GoI has allocated a sum of ₹ 5,300 crore for micro irrigation, watershed development under 'Pradhan Mantri Krishi Sinchai Yojana' (Union Budget, 2015-16).

- Increased allocation of Budget - The agriculture sector was allocated a budget of ₹ 47,912 crores for the year 2016-17, which is 84% more than the budget allocated for the year 2015-16. The total budgetary allocation for rural, agricultural and allied sectors is ₹ 187,223 crore in 2017-18. Union Budget 2020 benefits the Indian Agriculture Sector by the allocation of ₹ 2.83 lakh crores.

- Establishment of Mega Food Parks - The Ministry of Food Processing Industries (MoFPI) has undertaken an initiative to establish 42 mega food parks across the country, to increase the level of processed food from the 10% in 2015 to 20% in the next few years. Attractive incentives have also been introduced by the central and state governments to include capital subsidies, tax rebates, and reduced custom and excise duties. The central government is also encouraging disbursement of loans under a priority sector lending scheme to ensure that entrepreneurs have access to credit to set up food processing units.

- **Cold chain sector initiatives:**
  - GoI has recognised the cold chain industry as a sub-sector of infrastructure in the Union Budget in 2015 and investment in the cold chain has been opened under the automatic route for 100% FDI participation.
  - There has been viability gap funding of up to 40% of the cost.
  - 5% concession has been provided on import duty, service tax exemption and excise duty exemption on several items.
  - Subsidy of over 25 to 33.3% is provided for cold storage projects
  - The central government has approved 138 integrated cold chain projects. To develop the cold chain supply and to increase storage capacity, a National Cold Chain Development office has been established. The office of the Foreign Agricultural Service of the USA, in New Delhi has been collaborating with the Global Cold Chain Alliance (India chapter) to identify areas where US technology and expertise can help develop the cold chain sector in India.

- **Easier access to credit** - The government has helped arrange easier access to credit on behalf of farmers. These farmer-friendly policies and the new and growing trend of collaborative farming in India have encouraged the farming community to embrace mechanisation, leading to a structural shift in demand towards high-powered agricultural machineries. Some state governments, with support from the central government, have embarked on a public-private partnership (PPP) model to start custom hire centres to provide agricultural machinery on a rental basis to farmers.

- **The National Agriculture Market (NAM):** NAM was set up as a pan-India electronic trading portal to network existing APMC mandis. By the end of March 2017, 417 mandis across 13 states had been integrated with the e-NAM Portal. As of May 15, 2017, 83.57 lakh tonnes of agricultural produce worth ₹19,802.98 crores had been traded on the platform. Furthermore, 45,45,850 farmers, 89,934 traders and 46,411 commission agents had registered on the platform.

- Creation of a long-term irrigation fund - The government has created a dedicated long-term irrigation fund in NABARD with an initial corpus of about Rs. 20,000 crore and raised the agriculture credit target to Rs. 9 lakh crores for 2016-17 as against the target of Rs. 8.5 lakh crore during 2015-16. In 2017-18, target for agricultural credit has been fixed at Rs. 10 lakh crores.

- Introduction of the Krishi Kalyan Cess - In order to finance initiatives to improve the agriculture sector, the government has imposed a 'Krishi Kalyan Cess' of 0.5% on all taxable services.

- **Introduction of new projects and measures** – Some of the other measures that have been introduced to help farmers in increasing their income through farming and allied activities are:

- Pashudhan Sanjivani - Animal welfare programme and provision of animal healthcare cards.

- e-Pashudhan Haat - e-Market portal for connecting farmers and breeders; National Genomic Centre for indigenous breeds

- **Increase in the land to be brought under irrigation** – Budget 2016-17 proposed to bring around 2.85 million hectares of land under irrigation so as to improve agriculture and increase farmers welfare. The government has started work on 99 major and medium irrigation projects, slated to be completed by 2019.

- In Budget 2017-18, the Long Term Irrigation Fund already set up in NABARD has been proposed to be augmented by 100% to take its total corpus to Rs. 40,000 crores. Furthermore, a Dedicated Micro Irrigation Fund has been planned under NABARD to achieve 'per drop more crop' with an initial corpus of Rs 5,000 crore.

- **100% rural electrification** – GoI has planned for 100% village electrification by 1 May, 2018. Rural electrification is a must-have in rural areas as this helps in meeting various energy needs including basic lighting, irrigation, communication, water heating, etc. The availability of power will further improve farm productivity as a result of substitution from manual tools to power-based equipment.

- **Small Farmer Agribusiness Consortium (SFAC):** Department of Agriculture and Cooperation commissioned Small Farmers Agri-business Consortium (SFAC) in 2011-12 to promote the formation of Farmer Producer Companies which was considered as a key strategic goal under the

12th five-year plan. A farmer producer company is an amalgamation between a private company and a cooperative society which addresses the rising need of an institutional structure that consists of ethos of cooperation on one side and business resilience on the other side. Formation of FPCs leads to collectivization of farmers which helps in bringing down the input costs, enabling better bargaining power at the hands of the farmer and magnifies their voices. India currently has around 600 FPCs, of which a third are in Maharashtra alone. On January 1, 2014, SFAC launched a Central Sector Scheme "Equity Grant and Credit Guarantee Fund Scheme for Farmers Producer Companies" to support the Farmer Producers Organizations (FPOs) in terms of strengthening their capital base. The two main components of the schemes are:

- Equity Grant Scheme: Every registered Farmer Producer Company (which is registered under the special provision of the Companies Act) is provided a grant of upto ₹ 10 lakh with an objective to enhance the equity base of the FPC and enable it to approach financial institutions for raising working capital.

- Credit Guarantee Fund (CGF): The CGF offers a cover of 85% (up to a maximum of ₹ 1 crore) to loans extended by banks to Farmers Producer Companies without collateral.

The Population and Food Scenario

It is interesting to recall that the world population was 1 billion in 1800 AD; 2 billion in 1930 (1 billion being added in 130 years), 3 billion in 1960 (1 b in 30 years), 6 billion in 2000 (3 billion in only 40 years) and projected to be 10 billion+ in 2100 (4 billion in 100 years). Major contributions to population growth came from Africa, Asia and Latin America. Today, considering its size, India is among the densest regions in the world. By 2050 India's population will be about 1.7 billion which will be the highest in the world and about 400 million more than China- the most populous nation today. How to feed them and preserve our finite production and life support resource systems are issues of highest concern. There is no way to postpone or ignore these issues and challenges any longer.

The population change in India since 1950 and the UN's projections of population by age bracket. Here we see that the number of children under the age of five (under-5s) peaked in 2007; since then the number has been falling. The number of Indians less than 15 years old peaked slightly later (in 2011) and is now also declining. These are landmark moments in demographic change. India's population will still continue to grow as a result of 'population momentum' – the effect often referred to by Hans Rosling and Gap minder as the 'inevitable fill-up' when young generations grow older. But we can now see an end to population growth: reaching 'peak child' anticipates the later 'peak population'. The number of children has peaked; the total population will follow and reach its peak in four decades. In 1950, the Indian population was only 376.32 million which is expected to overtake China and become the most populous nation in 2024. It is estimated that by 2030, the Indian population would be 1.52 billion and approximately 1.70 billion by 2050. It is after 2060 that the population will stabilize and start declining.

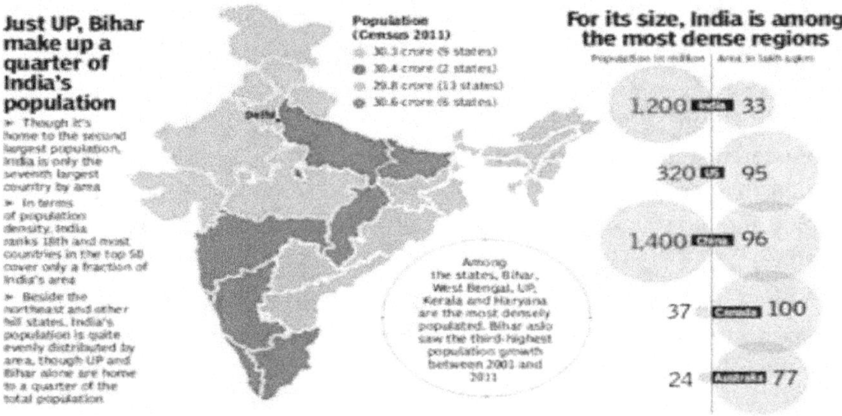

Figure 1.10 Population density in India

India has been able to feed its vast population and enacted the National Food Security Act which legally empowers the population below the poverty line to be provided with the basic food requirements. But with the population expected to reach 1.7 billion by 2050, the pressure on land, water and other resources to meet its food and development needs is going to be very intense. Food and nutritional security are also threatened by issues such as the severe decline in the health and productivity of the soil leading to a decline in Total Factor Productivity, low nutrient content in the food, poor health of the crops predisposing them to severe insect-pests and diseases, ultimately resulting in poor health of human beings and animals.

By 2050, India needs to step up production of all agricultural commodities by around 30 per cent in food grains and to more than 300 per cent in vegetable oils to meet the needs of increased population and rising living standards. To meet the expected higher demand for agricultural commodities, the most appropriate strategy would appear to be improvement of productivity through various measures, including increasing cropping intensity, adoption of the best water, soil and land care management practices, along with the preservation of prime agricultural lands, water resources and biodiversity.

1.4 Financial Services

Finance plays an important a role in supporting agriculture in tune with development of technologies. Liberal credit support by lending institutions in the past enabled rapid infrastructural growth and improved farm level credit absorption capacity. Availability of, and accessibility to, credit, has, however, shown acute skewness over regions, subsectors of the agriculture sector (such as minor irrigation or horticulture) and sections of the farming community such a small/marginal farmers, the youth and women, etc. Some of the states with better natural resource base have progressed well, while some others lagged far behind. Likewise, some farmers with better resource endowments and access to financial and other institutions have done better. Furthermore, the multiplicity of lending institutions and the liberal deployment of credit through various ongoing schemes including micro-financing have saved rural dwellers from the clutches of money lenders to a large extent.

The achievement of targets in the agricultural sector which covers the production of food and essential raw material like cotton, jute and oilseeds, ought not to be allowed to suffer for want of adequate credit. However, specific items of productive work and rates of interest need to be considered as an integral part of the Plan. For providing these facilities all the existing agencies like money lenders, Commercial Banks (CBs), cooperatives and the State have to be integrated and harnessed to a common purpose. Such a comprehensive approach is essential for ensuring the optimal use of all the available resources of the nation.

Classification of Finance

Short-Term: "Short-term loans" or "seasonal agricultural operations" loans are generally advanced for meeting annual recurring expenses such as, seed, feed, fertilizers, pesticides, weedicides hired labour expenses, and hired machinery charges which are termed as seasonal loans/crop loans/production loans. These are expected to be repaid after the harvest. It is expected that the loan together with interest will be repaid from the income received through the enterprise in which it was invested. The time limit to repay such loans is a year.

Medium-Term: "Medium-term loans" are advanced for comparatively longer lived assets such as machinery, diesel engine, wells, irrigation structure, threshers, shelters, crushers, draught and milch animals, dairy/poultry sheds, etc., where the returns accruing from an increase in farm assets are spread over more than one production period. The usual repayment period for such types of loan is from fifteen months to five years.

Long-Term Loans: Repayable over a longer period (i.e., above 5 years) are classified as long-term loans. "Long-term loans" are related to the long-life assets such as heavy machinery, land and its reclamation, erection of farm buildings, construction of permanent-drainage or irrigation system, etc. which require large sums of money for the initial investment. The benefits generated through such assets are spread over nearly the entire life of the asset. The normal repayment period for such loans ranges from five to fifteen or even up to 20 years (Desai. S.S.M, 1990).

Traders and Commission Agents: Traders and commission agents advance loans to agriculturists for productive purposes against their crop without completing legal formalities. It often becomes obligatory for farmers to buy inputs and sell outputs through them. They charge a hefty rate of interest on the loan and a commission on all the sales and purchases, making it exploitative in nature.

Landlords: Mostly small farmers and tenants depend on landlords for meeting their production and day to day financial requirements.

Money Lenders: Despite the rapid growth of rural branches of different institutional credit agencies, village moneylenders continue to have an important role. They are of two types, agriculturist moneylenders who combine their money lending jobs with farming and professional moneylenders whose sole job is money lending.

A number of reasons have been attributed for the popularity of moneylenders such as: they meet the demand for productive as well as unproductive requirements, they are easily approachable even at odd hours; and they require very low paper work and advances are given against promissory notes or land. Moneylenders charge a huge rate of interest as they take advantage of the urgency of the situation. Over the years a need for regulation of money lending has been felt. But lack of institutional credit access to certain sections and areas have facilitated the unhindered operation of money lending.

Institutional Credit Agencies: The evolution of institutional credit to agriculture could be broadly classified into four distinct phases - 1904-1969 (predominance of Cooperatives and setting up of RBI), 1969-1975 (Nationalization of CBs and setting up of Regional Rural Banks (RRBs)), 1975-1990 (setting up of NABARD) and from 1991 onwards (financial sector reforms). Institutional funding of the farm sector is mainly done by CBs, regional rural banks and cooperative banks. The share of CBs in total institutional credit to agriculture is almost 48 per cent followed by cooperative banks with a share of 46 per cent. Regional Rural Banks account for just about 6 per cent of total credit disbursement.

Cooperative Credit Societies: The history of the cooperative movement in India dates back to 1904 when the first Cooperative Credit Societies Act was passed by the Government. The scope of the Act was restricted to the establishment of primary credit societies and non-credit societies. The shortcomings of the Act were rectified through the passing of another Act called the Cooperative Societies Act 1912. The Act gave provision for registration of all types of Cooperative Societies. This made the emergence of rural cooperatives both in the credit and non-credit areas, though with uneven spatial growth. Soon after the independence, the GoI following the recommendations of the All India Rural Credit Survey Committee (1951) felt that cooperatives were the only alternative to promote agricultural credit and development of rural areas. Accordingly, cooperatives received substantial help in the provision of credit from the Reserve Bank of India as a part of loan policy and large scale assistance from Central and State Governments for their development and strengthening. Many schemes involving subsidies and concessions for the weaker sections were routed through cooperatives. As a result cooperative institutions registered remarkable growth in post-independent India.

Cooperatives play a very important role in the disbursement of agricultural credit. Credit is needed both by the distribution channel as well as by the farmers. The distribution channel needs it to finance the fertilizers business and farmers need it for meeting various needs for agricultural production including purchasing fertilizers. The credit needed by the farmers for the purchase of fertilizers and other inputs is called 'short term' credit or 'production credit' whereas, credit needed by the distribution channel is called 'Distribution Credit'. Cooperatives also play a very important role in the disbursement of 'Medium Term' and 'Long Term' credit needed by the farmers'. In India, 78 per cent of the farmers belong to the category of small and marginal farmers, and they depend heavily on credit for their agricultural operations. These farmers will not be able to adopt modern agricultural practices unless they are supported by a system that ensures adequate and timely availability of credit on reasonable terms and conditions.

Commercial Banks: Previously CBs were confined only to urban areas serving mainly the activities of trade, commerce and industry. The insignificant participation of CBs in rural lending was explained by the risky nature of agriculture due to its heavy dependence on monsoon, unorganized nature and subsistence approach. Following the nationalisation of CBs in 1969 and they played an active role in agricultural credit was accelerated, and they are the largest source of institutional credit to agriculture.

Regional Rural Banks (RRBs): RRBs were set up in those regions where availability of institutional credit was found to be inadequate but the potential for agricultural development was very high. However, the main thrust of the RRBs is to provide loans to small and marginal farmers, landless labourers and village artisans. These loans are advanced for productive purposes. At present 56 RRBs are functioning in India to support rural people, particularly to weaker sections.

Micro financing: Micro financing through Self Help Groups (SHG) has assumed prominence in recent years. SHGs is a group of rural poor who volunteer to organise themselves into a group for the eradication of poverty of the members. They agree to save regularly and convert their savings into a common fund known as the Group corpus. The members of the group agree to use this common fund and such other funds that they may receive as a group through common management. As soon as the SHG is formed and a couple of group meetings are held, an SHG can open a Savings Bank account with the nearest Commercial or Regional Rural Bank or a Cooperative Bank. This is essential to keep the thrift and other earnings of the SHG safely and also to improve the transparency levels of SHG's transactions. Opening of SB account is the beginning of a relationship between the bank and the SHG. Once this process is over, banks liberally lend to the groups or to members and recover the loans conveniently. The banks even offer a subsidy to the amount of loans borrowed based on their good response.

Micro Credit: Micro credit is the extension of small loans to entrepreneurs too poor to qualify for traditional bank loans. In developing countries, especially, micro credit enables very poor people to engage in self-employment projects that generate income. Though the micro credit financing has got its root from the development of Grameen Bank in Bangladesh in 1976s, and now it has been expanded globally, not only in terms of covering a larger number of clients but also with the increasing number of microfinance institutions that come forward to carry out these services for the poor. In order to give further fillip to micro-finance movement, the RBI has enabled Non-Governmental Organisations (NGOs) engaged in micro-finance activities to access external commercial borrowings (ECBs) up to US\$ 5 million during a financial year for permitted end-use, under the automatic route, as an additional channel of resource mobilisation. RBI is moving towards a systems perspective for providing effective policy support not only because a number of different institutions, viz., banks, MFIs, NGOs and SHGs are involved, but also because these institutions have very different institutional goals.

Non-Governmental Organisation (NGO): An NGO is a voluntary organization established to undertake social intermediation like organizing SHGs of micro entrepreneurs and entrusting them to banks for credit linkage of financial intermediation like borrowing bulk funds from banks for on-lending to SHGs. The microfinance sector has emerged from the efforts of Non-Governmental Organisations (NGOs), and as a response to the failure of existing structures to deliver financial services to the poor. The efforts by NGOs have emerged from grassroots and represent diversity.

Self-Help Group (SHG): An SHG is a registered or unregistered group of micro entrepreneurs having homogenous social and economic background voluntarily, coming together to save small amounts regularly, to mutually agree to contribute to a common fund and to meet their emergency needs on mutual help basis. The group members use collective wisdom and peer pressure to ensure proper end-use of credit and timely repayment thereof. In fact, peer pressure has been recognized as an effective substitute for collaterals. The RBI has also permitted all the banks as a special case to provide access to credit to unregistered SHGs.

Categories of the Priority Sector

Agriculture (Direct and Indirect finance): Direct finance to agriculture includes short, medium and long term loans given for agriculture and allied activities (dairy, fishery, piggery, poultry, beekeeping, etc.) directly to individual farmers, Self-Help Groups (SHGs) or Joint Liability Groups (JLGs) of individual farmers without limit and to others (such as corporate, partnership firms and institutions) for taking up agriculture/allied activities.

Small Enterprises (Direct and Indirect Finance): Direct finance to small enterprises include all loans given to micro and small (manufacturing) enterprises engaged in manufacture/ production, processing or preservation of goods, micro and small (service) enterprises engaged in providing or rendering of services, and whose investment in plant and machinery and equipment (original cost excluding land and building and such items as mentioned therein). The micro and small (service) enterprises include small road and water transport operators, small business, professional and self-employed persons, and all other service enterprises. Indirect finance to small enterprises include finance to any person providing inputs to or marketing the output of artisans, village and cottage industries, handlooms and to cooperatives of producers in this sector.

Retail Trade: includes retail traders/private retail traders dealing in essential commodities (fair price shops), and consumer cooperative stores.

Micro Credit: Provision of credit and other financial services and products of very small amounts not exceeding ₹ 50,000 per borrower, either directly or indirectly through a SHG/JLG mechanism or to NBFC/MFI for on lending up to ₹ 50,000 per borrower, will constitute micro credit.

Crop Insurance

Crop insurance is availed of by farmers, ranchers, and others to protect themselves against either the loss of their crops due to natural disasters, such as hail, drought, and floods, or the loss of revenue due to declines in the prices of agricultural commodities.

The Crop insurance schemes aim at providing comprehensive risk insurance which covers the yield losses that occur to the agricultural output of small and marginal farmers due to non-preventable risks.

The crop insurance risks covered under the non-preventable category are listed below:

- Natural Fire and Lightning' Storm, Hailstorm, Cyclone, Typhoon, Tempest, Hurricane and Tornado
- Flood, Inundation and Landslide
- Drought and Dry spells
- Insect Pests and Diseases

The sum insured under the crop insurance risks covered usually extends to the value of the threshold yield of the insured crop. This is usually subject to the option of the insured farmers. Nevertheless, a farmer may also choose to insure his crop beyond the value of the threshold yield level up to 150 per cent of average yield of the notified area on payment of premium at commercial rates. Apart from the risks covered in the crop insurance scheme, what is important is the sum insured. In case of Loanee farmers the sum insured would be at least equal to the amount of crop loan advanced. Further, in the case of the Loanee farmers, the insurance charges that will be levied will be additional to the Scale of Finance for the purpose of obtaining loan. Apart from the above-mentioned issues, the matters of Crop Loan disbursement procedures, which have been outlined by the RBI or NABARD are binding. The insurance premium issues still stand at an undecided state as the transition to the actuarial regime in case of cereals, millets, pulses and oilseeds is expected to be made in a period of five years.

The government through its various schemes of finance to agriculture aims helping farmers and thereby promoting the growth and productivity of agriculture. This enhancing attempt is a major encouragement to farmers to sustain the growth in agriculture. The schemes of finance explained is a continuous and sustained effort of the government to encourage farmers.

The Crop Insurance Schemes

Crop insurance is an important measure undertaken by the central government in India to manage the risk associated with drought, unseasonal rainfall, floods, other natural calamities. The most popular scheme in this area is Pradhan Mantri Fasal Bhima Yojana (PMFBY). Based on past experience, PMFBY was introduced by the central government in 2016-17 with the objective of covering the total cropped area in the country through the modest yearly target of covering 30 per cent of the cropped area. In 2016-17 the target was achieved.

Market determined premia are fixed, based on an open tendering system for each cluster comprising a few districts (2-5 neighbouring districts in one cluster) with similar agro-ecological conditions. In this, all insurers (insurance companies) can participate in the bidding and the lowest bidder in terms of premium quoted. The final selection of the successful bidder for the cluster is based on the lowest weighted average premium rate. All crop insurance activities in that particular cluster have to be operated by the winner- insurer for the notified year. Under this scheme, there are about 5 public and 13 private insurers who bid for the cluster for all notified crops.

Evidence from many states suggest, that there are insufficient bidders in rain fed, remote and backward districts. This could result in an oligopoly situation under which there is scope for cartelisation of the insurers to rig premium rates, so that market determined premium rates are higher than the actuarial rates. In these backward districts, it was found that in some instances market determined premiums were as high as 25 per cent due to few bidders participating in the bidding process, leading to above normal profits for the insurers.

The government needs to take steps to increase competition among insurers to lower market-determined premium rates charged by insurers. Further, there is a need to strengthen the State Level Co-ordination Committee on Crop Insurance (SLCCCI) and District Level Monitoring Committee (DLMC) in collaboration with financial institutions and insurers to promote healthy competition.

In India, most farmers are unaware of PMFBY and its modalities. First, there is a need to increase awareness in frontline field staff like, department of agriculture, banks and insurers, who could be the game changers for the scheme. Currently, they themselves need to be educated about the scheme modalities which will help them in disseminating proper information to farmers. There is a need to sensitize the first contact point of farmers like input dealers, Gram Panchayat members and Agricultural Extension Officers (AEOs) to increase awareness of the scheme.

Uniform cut-off dates for premium payments are not ideal for diverse agro-climatic conditions across the districts. There needs to be flexibility in deciding the cut-off dates for premium payments at the local level. This task could be entrusted to the DLMC. The roles and responsibilities of DLMC needs to be enhanced so that, it can announce cut-off dates in time, conduct field level inspections in case of any calamity and decide mid-season or end of the season claims and see that every farmer gets it in time. The denial and delay of claim issue can be solved, that can reduce dependency on money lenders.

There is also inadequate infrastructure in terms of dedicated field staff of the insurers to address grievances of the farmers, to inspect yield losses etc. Therefore, there is need to increase the efficiency of the whole set of operations like the functioning of weather stations, conduct of crop-cutting experiments, etc., by measures such as the use of drones, satellite images, and GIS-based mobile phones.

Agriculture Inputs

Indian agriculture has consistently struggled with low productivity. Thus, agricultural inputs are a key sector of focus to tackle the challenge of low yield and support the Doubling Farmers' Income initiative.

Seed Market

India has a share of 4% in the global seed market. The Indian seed market reached a value of US$3.6 billion in 2017, exhibiting a CAGR of about 17% during 2010-2017. The seed market is expected to surpass US$8 billion by 2023.25 Indian seed market is dominated by maize, cotton, paddy, wheat, sorghum, sunflower and millets.

India, being a major agriculture-based economy, is also a key player in the agricultural seed trade. The sector shows huge potential for growth supported by seed policies in favour of seed producers and exporters. The seed industry is exempted from the Goods and Service Tax (GST) rate on seeds used for sowing purposes only. The Government has also allowed 100% Foreign Direct Investment (FDI) in the development and production of seeds and planting materials.

Strategic Interventions

Setting up seed production hubs: Currently, there are no dedicated seed production hubs in India. Setting up dedicated seed production hubs can lead to focused efforts in sector development. Inclusion of the private sector is necessary for the successful operation of these hubs, as it can bring in the necessary expertise to the sector. Government incentives for investment in these hubs can help attract private players.

Export orientation: As mentioned, India has a huge scope for the promotion of seed exports and requires efforts in this area. The seed production hubs can be oriented towards exports by establishing an export promotion wing. The functions of this wing may include global seed trade assessment, thorough understanding and tracking of seed import requirements across geographies, and consequently, alignment of the seed production as per import requirements.

Strengthening of Seed Village Programme: Strengthening the Seed Village Programme will improve seed delivery. Seed subsidies for high yielding variety and hybrid seeds may be given to farmers at specified intervals, so that the maximum number of farmers gain access to newly developed seeds.

Seed cost rationalisation: In the present system for vegetable hybrids, the varieties and hybrids developed by ICAR-SAUs are given to private sector seed companies at nominal rates of royalty. By providing breeder seeds on a cost basis to public sector seed organisations, they may be able to produce vegetable seeds on a large scale and provide them to farmers at reasonable prices. ICAR and SAUs should also provide variety wise packages of practices for vegetable seed production and technology, along with breeder seeds.

Improving seed traceability: To control the distribution of spurious seeds, improving seed traceability is an important measure that needs to be in place. Using a barcode or Quick Response (QR) Code could be one method. Using these unique codes could help in strengthening seed traceability and enable farmers to track the origin of the seed before purchasing. Besides deterring spurious seeds, it could also help in tracing their origin in case of quality issues.

Agrochemical Industry

The agrochemical industry is also integral to Indian agriculture. The agrochemical industry is comprised of insecticides, fungicides, herbicides, bio-pesticides and other chemicals, such as fumigants and plant growth regulators. The Indian agrochemical market is pegged to reach US$6.3 billion by FY20.Further, at the global level, India remains one of the lowest users of agrochemicals even though it is a major food producer. While India ranks second in global food production, its share in agrochemical usage is comparatively very low, ranking 13 globally.

The rationalisation of pesticide regulation is necessary to reduce the time taken to register new molecules and create a more efficient system. Currently, most new molecules with patents are imported. Impetus on the domestic production of new molecules by incentivising such efforts is necessary. Currently, the organic pesticides and bio-stimulants category is not covered under the Pesticides Act, and thus, a comprehensive policy regulation is necessary.

Fertilizers

Given the heavy dependency of our economy and employment on agriculture, the fertiliser sector continues to be a major focus area for policymakers. Fertilisers played a significant role in the Green Revolution and overall agricultural growth in the country. With the increasing demand for production, usage and growth in the fertiliser sector have also seen an upward trend. However, India still lags behind in average fertiliser usage when compared to developed economies. There are also significant variations in inter-country usage.

To prevent urea diversion from agriculture, the Department of Fertilizers has mandated that 100% of production by all domestic manufacturers should be neem-coated urea. Currently, most of the fertiliser usage is not based on the soil testing report and overlooks the location and crop-specific nutrient status. This may impact the yield levels of the crop. The Soil Health Card Scheme is one of the positive initiatives by the Government in this direction, although implementation challenges still persist. Linking of Soil Health Cards to other measures like agri finance and subsidies need to be taken up on priority.

Despite the increase in fertiliser use in the last few decades, food grain production has not grown proportionately. Today, we are producing less food per kg of Fertiliser than we were earlier. The decline in fertiliser use efficiency is attributed to dwindling organic carbon in Indian soils, micronutrient deficiencies and imbalanced fertiliser use. The country lags behind in the introduction of high efficiency fertiliser technologies capable of reducing wastage, like nanotechnology, nitrification inhibitors and urease inhibitors which are popular in other parts of the world.

Implements/ Farm Mechanisation

India accounts for around one-third of the global tractor production. However, the level of farm mechanisation in India is still lagging behind developed economies. With the Government's increasing focus on farmer remuneration, it has become imperative to increase the use of farm machinery and farm equipment, especially with regard to irrigation practices. Custom hire centres have been developed by many states on a PPP basis to rent out agricultural equipment. The level of mechanisation in India stands at about 40-45%.

There is a high level of regional variation in the level of mechanisation in India. States like Punjab, Haryana and Uttar Pradesh showcase high mechanisation in agriculture, while states in the south, west and northeast lag behind. The high level of mechanisation may be attributed to favourable government policies, large landholdings and high awareness. The Indian farm mechanisation market is expected to reach ₹ 400 billion by 2019-20.

Access to finance for farm equipment must be improved by simplifying the documentation process and providing interest subvention for timely payments. Most of the subsidy released is a back-ended producer subsidy and is often delayed. The process needs to be converted to Direct Benefit Transfer (DBT) to ensure timely and rightful subsidy release. Custom Hiring Centres (CHCs) tackle many challenges in farm mechanisation. However, they face certain challenges themselves, such as availability of farm machinery and implements during peak period and underutilisation during lean period, lack of awareness and support services. To improve the efficiency of CHCs, certain measures could be adopted, such as in addition to providing access to machines on rental providing access to a package of practices, quality inputs, extension services and input application equipment.

Production Scenario

As per Fourth Advance Estimates for 2019-20 released on 19.08.2020, total food grain production in the country is estimated 296.65 million tonnes. Total production of rice during 2019-20 is at a record 118.43 million tonnes, production of wheat, estimated at record 107.59 million tonnes, pulse production 23.15 million tonnes, oils seeds production of nine seeds (groundnut, castor, sesamum, niger seed, soya bean, sunflower, rapeseed, mustard, linseed, safflower) are 334.23 million tonnes. Total production of sugarcane in the country during 2019-20 is estimated at 3557.00 million tonnes, cotton as 354.91lakh bales of 170 kg each, total jute and mesta estimated at 99.06 lakh bales of 180 kg each.

The total production of rice during 2018-19 is at a record of 116.48 million tonnes. Production of rice has increased by 3.66 million tonnes than the production of 112.76 million tonnes during 2017-18. Production of wheat, estimated at record 102.19 million tonnes, is higher by 2.32 million tonnes as compared to wheat production of 99.87 million tonnes achieved during 2017-18. Total pulses production during 2018-19 is estimated at 23.40 million tonnes. Total oilseeds production in the country during 2018- 19 is estimated at 32.26 million tonnes which is higher than the production of 31.46 million tonnes during 2017-18. With an increase of20.25 million tonnes over 2017-18, the total production of sugarcane in the country during 2018-19 is estimated at 400.16 million tonnes. Production of cotton estimated at 28.71 lakh bales (of 170 kg each) and production of jute and mesta estimated at 9.77 lakh bales (of 180 kg each).

Irrigation in India

As agriculture is the main water consuming sector in India, there is a need for sound water-management in water-scarce regions. The Irrigation sector currently consumes 80% of the total water used. Owing to competing demands from other sectors, it is expected that water consumption in this sector will probably reduce to about 70% by 2050. According to a World Bank report, groundwater has supported 60% of irrigated agriculture, whereas 40% of irrigated agriculture is supported by surface water. This excessive dependence on groundwater, has, in many areas, caused over-extraction of groundwater, an issue that needs to be addressed.

The National Water Policy (NWP), 2012 states that water saving in irrigation is of utmost importance. Goal 4 of the National Water Mission (NWM), 2008 of India highlights the main objective of NWM, which is to improve water use efficiency by at least 20% in all sectors, including domestic, industrial, agricultural and commercial. This objective can be attained by enhancing the efficiency of the demand side and the supply side in the agriculture sector by the use of micro irrigation techniques. Below given table represents the source-wise net irrigated area and the percentage of the net irrigated area by source (in million hectares) in India. As per the table, from 1960 to 2015, the area under canal irrigation increased from 10.37 Mha (million hectares) to 16.8 Mha, whereas the area under tube-well irrigation increased from 0.13 Mha to 31.60 Mha, and overall there had been an increase in the net irrigated area from 24.66 Mha to 68.38 Mha. The increase in net irrigated area as well as the shift in dependence from surface water to groundwater has had a detrimental impact on the groundwater resources. As water resources become increasingly scarce, there will arise the need to manage irrigation water efficiently and the adoption of micro irrigation systems.

Table 1.13 Ten years trend on source-wise net irrigated area and percentage wise net-irrigated area in India.

	Source-wise net irrigated area and the percentage of the net irrigated area by source								
Year	Canal		Tanks		Tube wells		Other wells		Net irrigated area
	Mha	%	Mha	%	Mha	%	Mha	%	Mha
1960/61	10.37	42.05	4.56	18.49	0.13	0.55	7.15	29.01	24.66
1970/71	12.83	41.28	4.11	13.22	4.46	14.34	7.42	23.88	31.10
1980/81	15.29	39.49	3.18	8.22	9.53	24.62	8.16	21.08	38.72
1990/91	17.45	36.34	2.94	6.13	14.25	29.62	10.43	21.73	48.02
1995/96	17.12	32.06	3.11	5.84	17.89	33.51	11.80	22.10	53.40
2000/01	15.71	28.65	2.51	4.59	22.32	40.71	11.45	20.88	54.83
2005/06	16.72	27.50	2.08	3.40	26.03	42.80	10.04	16.50	60.84
2010/11	15.64	24.6	1.98	3.10	28.54	44.8	10.63	16.70	63.66
2013/14	16.27	23.90	1.84	2.70	31.13	45.70	11.31	16.60	68.10
2014/15	16.18	23.66	1.72	2.52	31.60	46.21	11.35	16.60	68.38

Source: DES, MoA & FW, GoI

However, even though the overall potential micro irrigation in India is projected to be about 70 Mha (million hectares), actual coverage by 2018 was only around 9 Mha. Therefore, at the current coverage rate of 0.6 Mha/annum, it would take approximately more than 100 years to achieve the potential target of micro irrigation in India.

System of Water for Agriculture Rejuvenation (SWAR)

To drastically save and efficiently use water, the Centre for Environment Concerns, an NGO, has developed a unique, and first of its kind, an irrigation system of "delivering measured moisture at the plant root zone" called the System of Water for Agriculture Rejuvenation (SWAR). Sri K S Gopal developed SWAR won global champion water innovation awards in Paris and Washington and listed by NITI Aayog as a best practice in irrigation water use. SWAR is transformative to shift irrigation from measuring water and given on the surface to subsurface root zone area and based on moisture level and shows 40% water savings compared to drip systems by a scientific study and farmers' experience. Technically, SWAR involves storing of water in overhead tanks and sending it through a small diameter pipe to a customised locally made clay pot that is buried near the root area. The clay pot contains micro-tubes that transmit water through a sand pouch, to prevent the roots from invading the pipes and the pot. The slow oozing of water provides moisture for a prolonged period, the level of which is calculated based on soil type, plant species and their age. Thus, SWAR uses less water and wastes nothing. SWAR is used extensively for horticulture and forestry crops and comes as low-cost affordable add-on drip laterals.

Micro Irrigation

Micro irrigation ensures conservation and the efficient use of water, minimal wastage of water and higher productivity of crops with less water consumption by the usage of the drip irrigation method and the sprinkler irrigation method, respectively. For optimal and efficient use of surface and groundwater sources for irrigation, micro irrigation is one of the most-effective ways.

Micro irrigation includes the usage of drip and sprinkler systems. Micro irrigation could be one of the solutions to the challenges and issues faced by Indian agriculture. The water use efficiency of the flood method of irrigation in India is estimated to be only around 40%. This is mainly due to the significant losses through conveyance, distribution and evaporation, whereas, micro irrigation systems can provide water use efficiency from 80% to 95%. The reason for this difference is because transmission loss is nominal, while losses through evaporation, run-off and deep percolation are also reduced significantly by using micro irrigation methods. Efficient water use results in additional benefits such as an increase in the area coverage under irrigation with the same amount of water as well as increasing the potential usage of marginal/degraded land using micro irrigation systems.

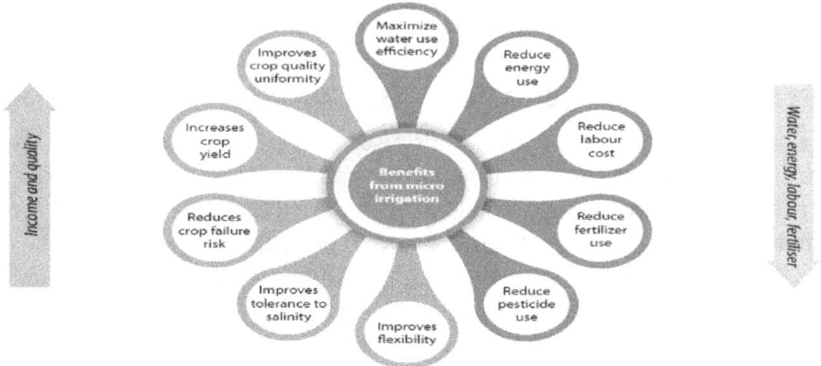

Figure 1.11 Benefits of micro irrigation adaptation

Source: Author compilation

The Indian Council of Food and Agriculture (ICFA) has established that farmers have an increase in income ranging from 24.5% to 70.5%, with an average increase in income of about 46.8% after micro irrigation systems adoption. The net-income increases following the adoption of micro irrigation systems need to be widely publicized so that these benefits can be availed on a large scale.

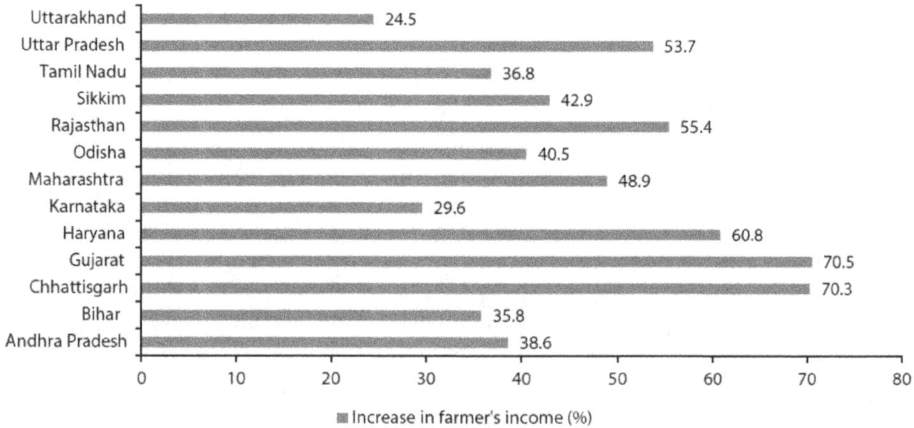

Figure 1.12 Increase in farmer's income

Source: Details available at https://icfa.org.in/assets/doc/reports/indian-micro-irrigation-market.pdf, last accessed on 15 May 2019

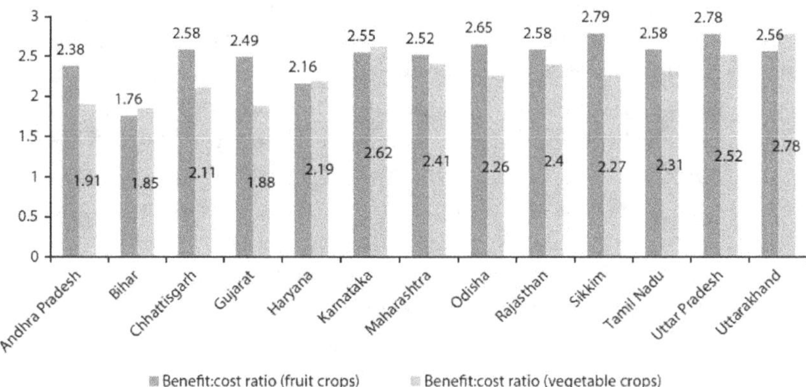

Figure 1.13 Benefit-cost ratio

Source: Details available at: https://icfa.org.in/assets/doc/reports/indian-micro-irrigation-market.pdf, last accessed on 15 May 2019

The average benefit cost ratio for the crops grown in each state indicates the benefits to the farmers of adopting micro irrigation system (MIS). The benefit-cost ratio also shows that horticulture crops seem to be more profitable as compared to vegetable crops in the majority of states such as Andhra Pradesh, Maharashtra, Gujarat, Odisha and Sikkim, where farmers have adopted MIS.

Agriculture Debt Waivers, WriteOffs and Subsidies

From ancient times, farming has been more a way of life than a commercial activity. If farming has proved to be sustainable, (leaving apart man-made famines like those in Bengal). In the exceptional cases of crop failures, since the British rule, compensation was available on an 'annavari' basis, i.e., if crop loss was more than 50-60 percent (eight annas in a rupee consisting of 16 annas). The concession was extended by way of postponement of repayment of principal or instalments, that too only for those who repaid outstanding interest. Further, these concessions were met out of mandated "Relief and Guarantee Funds" built up in each Cooperative Bank - the mainstay of Agriculture Credit (covering 80% farmers, mostly small and marginal farmers). In certain extreme cases, 'taccavi' loans were also given by governments directly to the identified affected farmers. It's worth noting that government support was not routed through the banking system to ensure no damage is done to the repayment ethics in the institutional credit channels.

While the system was going on systematically, post-independence- politics overlook good sense and the whole system got derailed. Pre-reforms, RBI/NABARD rigorously controlled/ stopped the unethical waivers and writeoffs. Some aggrieved states approached the Courts questioning the authority of the regulators in interfering with the States' actions. But to no avail. The NTR Govt. in AP in 1980s complied with the mandate to keep away Government support from institutional credit channels by choosing Panchayati Raj institutions. Overtime, both the Centre and State Govt. defied the regulators by liberally using the waivers and write-off route for political gains. It became a soft (though grossly wrong) option for politicians to paint indebtedness credit alone as the villain of distress in agriculture, ignoring other factors such as input - supply and markets, besides infrastructure support. That started the decline and fall of credit discipline and vitiated the repayment atmosphere of leading to the decay of institutions. "Financial Inclusion" took a back seat and the traditional money lender regained his supremacy.

The pleas of the intelligentsia to stop indiscriminate and across-the-board dole outs proved unacceptable in a democracy. If the ₹ 2 lakh crores of subsidies in the agriculture sector – are diverted to meet the requirement of rural infrastructure, the decline in the Agri sector could have been stopped and enduring growth could have been put back on track. "Cooperatives failed but cooperatives must succeed" – we need to wait and see if our democracy respects the sound appeal of eminent guardians of the Cooperative Movement, the hope for adequate, timely and effective credit to farmers.

Empowering the Small Farmers

Contributions of small holders in securing food for the growing population have increased considerably even though they are the most insecure and vulnerable group in society. Off-farm and non-farm employment opportunities can play an important role. Against expectation under the liberalized scenario, the non-agricultural employment in rural areas has not improved. Greater emphasis needs to be placed on non-farm employment and adequate budgetary allocations and rural credit through banking systems should be in place to promote appropriate rural enterprises. Specific human resource and skill development programmes to train them will make them better decision-makers and highly productive. Human resource development for increasing productivity of these small holders should get high priority. Thus, knowledge and skill development of rural people both in agriculture and non-agriculture sectors is essential for achieving economic and social goals. A careful balance will therefore need to be maintained between the agricultural and non-agricultural employment and farm and non-farm economy, as the two sectors are closely interconnected.

Raising agricultural productivity requires continuing investments in human resource development, agricultural research and development, improved information and extension, market, roads and related infrastructure development and efficient small-scale, farmer-controlled irrigation technologies, and custom hiring services. Such investments would give small farmers the options and flexibility to adjust and respond to market conditions.

For poor farm-households whose major endowment is its labour force, economic growth with equity will give increased entitlement by offering favourable markets for its products and more employment opportunities. Economic growth if not managed suitably, can lead to growing inequalities. Agrarian reforms to alleviate unequal access to land, compounded by unequal access to water, credit, knowledge and markets, have not only rectified income distribution but also resulted in sharp increases in productivity and hence need to be adopted widely. Further, targeted measures that not only address the immediate food and health care requirements of disadvantaged groups, but also provide them with developmental means, like access to inputs, infrastructure, services and most important, education should be taken.

CHAPTER - 2

CHALLENGES AND OPPORTUNITIES

Climate change presents a major threat to world agriculture affecting production, resources and livelihood. As many as 160,000 people will die every year in India by 2050 due to decreased food production because of climate change. India ranks second in the mortality forecast after China, whereas many as 248,000 are expected to die for this reason. The excess number of deaths annually due to heat in south Asia would be 21,648 in 2030 and 62,821 in 2050. Surprisingly, the US ranks fifth, after Vietnam and Bangladesh. By 2050, in the 155 countries, climate change will lower people's availability of food by 3.2%, fruits and vegetables by 4% and red meat by 0.7%. "Twice as many climate-related deaths were associated with reductions in fruit and vegetable consumption than with climate-related increases in the prevalence of underweight, and most climate-related deaths were projected to occur in South and East Asia.

Climate change affects the social and environmental determinants of health – clean air, safe drinking water, sufficient food and secure shelter. Between 2030 and 2050, climate change is expected to cause approximately 250 000 additional deaths per year, from malnutrition, malaria, diarrhoea and heat stress. The direct damage costs to health (i.e., excluding costs in health-determining sectors such as agriculture and water and sanitation), is estimated to be between $ 2-4 billion/year by 2030. Areas with weak health infrastructure – mostly in developing countries – will be the least able to cope without assistance to prepare and respond. Reducing emissions of greenhouse gases through better transport, food and energy-use choices can result in improved health, particularly through reduced air pollution.

It is estimated that by 2050, because of climate change, an additional 120 million people will be at risk of undernourishment. Further, global mean crop yields of rice, maize and wheat are projected to decrease 3% to 10% per degree of warming and impact livestock through reduced feed quantity/ quality, pest and disease prevalence, physical stress.

Climate change is a harsh reality, and is already evidenced in a shift observed in some cropping patterns. Climate has a direct impact on cultivation, including farm profitability and the long-term. Changes in weather patterns will also affect associated food security.

Source: Photograph by the author

Agriculture continues to be fundamentally dependent on the weather and will remain sensitive to short-term variations in weather and to seasonal, annual and long-term changes in climate. Climate change is not just the warming of air temperature, but the linked long-term alteration in established weather patterns. The change manifests initially in weather disruptions such as unseasonal rains, winds, floods, droughts, extreme warming or cooling and other incidents. Over the long run, it can cause a drastic shift in the agro-ecology with flora and fauna forced to adjust their life cycles or turn extinct.

Agricultural extension is that it conveys knowledge to farmers on how to manage their enterprises, increase productivity and raise their standard of living. Knowledge sharing in farming and rural enterprises goes back to the beginnings of agriculture and was characterised through the exchange of crop varieties and animal breeds as well as improved cultural practices.

In order to achieve the gigantic target of doubling farmers' income by 2022, there is an urgent need to address some of the basic challenges like climate change in the agriculture sector, disaster management for building a sustainable future, management of the agriculture sector in COVID-19 lockdown period with short, medium, long-term recommendations, natural resource management, precision agriculture and smart farming, research and extension services.

2.1 Climate Change

Agricultural production is being adversely affected by various factors such as rising temperatures, increased temperature variability, changes in levels and frequency of precipitation, greater frequency of dry spells and droughts, increasing intensity of extreme weather events, melting glaciers, rising sea levels, and the salinization of arable land and freshwater. These features of climate change will impact agriculture and it will soon become increasingly difficult to grow crops, raise animals, catch fish and manage forests.

The crops that we grow for food, fibre and energy need specific conditions in order to thrive, including optimal temperature and sufficient water. Up to a certain point, warmer temperatures may benefit the growth of certain crops. However, if temperatures exceed a crop's optimal level, or if sufficient water and nutrients are not available, yields are likely to fall. Increased frequency of extreme events, especially floods and droughts, also harms crops and reduces yields. Dealing with droughts could become a major challenge in areas where average temperatures are expected to increase and precipitation to decrease. Many weeds, insect pests and diseases thrive under warmer temperatures, wetter climates and increased levels of atmospheric carbon dioxide. Extreme temperatures, combined with decreasing rainfall, can prevent crops from growing altogether.

Heat waves, which are often caused by climate change, directly threaten livestock. Over time, heat stress increases animals' vulnerability to disease, thereby reducing fertility and meat and milk production. Climate change is also likely to modify the prevalence of livestock parasites and diseases. In areas with abundant rainfall, moisture-reliant pathogens thrive. Climate change also threatens the carrying capacity of grasslands and rangelands as well as feed production for non-grazing systems.

Global Warming

Global warming is the rising average temperature of Earth's atmosphere and since 1971, (Ninety) per cent of the warming has occurred in the oceans. Despite the oceans' dominant role in energy storage, and the term "global warming" is also used to refer to an increase in the average temperature of the air and sea at Earth's surface. Since the early 20th century, Earth's average surface temperature has increased by about $0.8°C$ ($1.4°F$), with about two-thirds of the increase occurring since 1980.Most of it is now accepted as being caused by increasing concentrations of Green House Gases (GHGs) produced by human activities such as deforestation and the burning of fossil fuels. According to the Fourth Assessment Report (AR4) of the Intergovernmental Panel on Climate Change (IPCC) the global surface temperature is likely to rise a forth 1.1 to $2.9°C$ (2 to $5.2°F$) for their lowest emissions scenario and 2.4 to $6.4°C$ (4.3 to $11.5°F$) for their highest. Other likely effects of the warming include heat waves, droughts and heavy rainfall, species extinctions on account of shifting temperature regimes, and changes in crop yields. The ecosystem services upon which human livelihoods depend would not be preserved because of the impact on the environment. Most countries are parties to the United Nations Framework Convention on Climate Change (UNFCCC), whose ultimate objective is to prevent "dangerous" anthropogenic (i.e., human induced) climate change, have adopted a range of policies designed to reduce greenhouse gas emissions and to assist in adaptation to global warming.

Glacial Melt

A glacier is a large persistent body of ice that forms where the accumulation of snow exceeds its ablation (melting and sublimation) over many years. Glacial ice is the largest reservoir of freshwater on earth, supporting one-third of the world's population. Glaciers store water and release as melt water, a source especially important for plants, animals and human uses. Glacial mass is affected by long-term climate changes and is considered among the most sensitive indicators of climate change and a major source of variations in sea level. Since the 1850s the retreat of glaciers has been affecting the availability of freshwater for irrigation and domestic use, mountain recreation, animals and plants that depend on glacier-melt, and in the longer term, the level of the oceans.

Sea Level Rise

Global average sea level rose at an average rate of around 1.7 + 0.3 mm per year from1950 to 2009. The two main factors which contributed to the rise are thermal expansion and the contribution of land-based ice due to increased melting. Current sea level rise can potentially impact human populations (e.g., those living in coastal regions and on islands and the natural environment (e.g., marine ecosystems).Sea level rise supports the view that the climate has recently warmed. IPCC has projected that during the 21st century, sea level will rise another 18 to 50 cm (7.1 to 23 in)

Chlorofluorocarbons (CFCs)

Chlorofluorocarbons (CFCs), have been implicated in the depletion of ozone in the Earth's stratosphere. These were developed in the early 1930s and are used in a variety of industrial, commercial, and household applications. Production and use of Chlorofluorocarbons experienced practically uninterrupted growth as demand for products requiring their use continued to rise. Not until 1973 was chlorine found to be a catalytic agent in ozone destruction. The announcement of polar ozone depletion over Antarctica in March 1985 prompted scientific initiatives to discover the Ozone Depletion Process, along with calls to freeze or diminish the production of chlorinated fluorocarbons. Global monitoring of ozone levels from space by the Total Ozone Mapping Spectrometer (TOMS) instrument has shown statically significant downward trends in ozone at all latitudes outside the tropics. Despite the rapid phase-out of CFCs, ozone levels are expected to be lower than pre-depletion levels for several decades due to the long tropospheric lifetimes of CFCs. Replacement compounds for CFCs have also been evaluated for their Ozone Depletion Potential (ODP).

Ozone Depletion

Ozone depletion describes two phenomena – a steady decline of about 4% per decade in the total volume of ozone in Earth's stratosphere (the ozone layer), and a springtime decrease in stratosphere ozone over Earth's Polar Regions. The latter phenomenon is referred to as the ozone hole. The most important cause is the catalytic destruction of ozone by atomic halogens. CFCs and other contributory substances are referred to as ozone-depleting substances (ODSs). The ozone layer prevents most harmful UVB wavelengths (280-315 nm) of ultraviolet light (UV light) from passing through the Earth's atmosphere. Therefore, decreases in ozone have generated worldwide concern leading to the adoption of the Montreal Protocol that bans the production of CFCs, halons, and other ozone-depleting chemicals such as carbon tetrachloride and trichloro-ethane. It is suspected that a variety of biological consequences such as increases in skin cancer, cataracts, damage to plants, and reduction of plankton populations in the ocean's photic zone may result from the increased UV exposure due to ozone depletion.

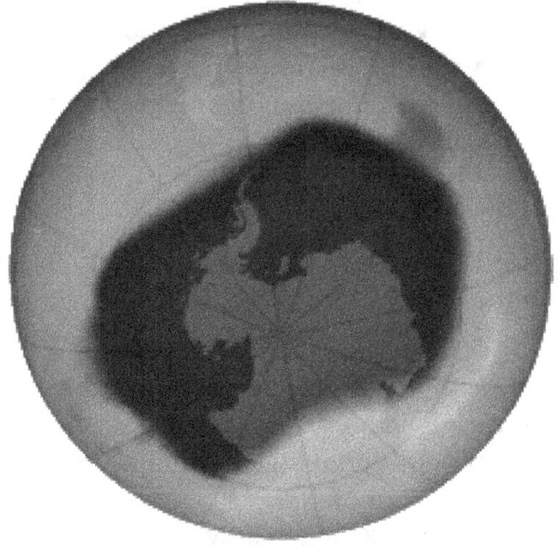

Renewable Energy

Renewable energy comes from natural resources - sunlight, wind, rain, tides, and geothermal heat. Climate change concerns, coupled with high oil prices, and increasing government support, are driving renewable energy legislation, incentives and commercialization. About 16% of global final energy consumption comes from renewables, 10% coming from traditional biomass, mainly used for heating, and 3.4% from hydroelectricity. New renewables (small hydro, modern biomass, wind, solar, geothermal, and biofuels) account for another 3% - growing rapidly. The share of renewables generation is around 19%, whereas 16% of global electricity coming from hydroelectricity and 3% from new renewables.

India and Climate Change

India and other developing countries feel strongly that they are not responsible for the threat of climate change and that rich industrialized nations in the world are responsible for it. Yet, the economies of these may be highly vulnerable. Food production would be adversely affected. Impacts of climate change could hinder development and delay progress in eradicating poverty, potentially aggravating social and environmental conditions. Precisely at a time when India is confronted with development imperatives, it is also likely to be severely impacted by climate change. India may face a major threat and serious adaptive capacity needs to be developed. Also, as in the food security or the global financial crises, it is in India's interest to ensure that the world moves towards a low carbon future. Impacts are already being seen in unprecedented heat waves, cyclones, floods, salinization of the coastline and effects on agriculture, fisheries and health. With a government target of 8% GDP to achieve developmental priorities, a share of one-sixth of the global population, and changing consumption patterns, India's emissions are set to increase dramatically. Although current per-capita emissions are among the lowest in the world, the IPCC report suggests that India will experience the greatest increase in energy and greenhouse gas emissions in the world.

Carbon Sequestration

Carbon sequestration and reductions in greenhouse gas emissions can occur through a variety of agricultural practices.

Climate change may have beneficial as well as detrimental consequences for agriculture. Some research indicates that warmer temperatures lengthen growing seasons and increased carbon dioxide in the air results in higher yields from some crops. A warming climate and decreasing soil moisture can also result in production patterns shifting northward and an increasing need for irrigation. Changes, however, will vary significantly by region. Geography will play a large role in how agriculture might benefit from climate change. While projections look favourable for some areas, the potential of increased climate variability and extremes are not necessarily considered. Benefits to agriculture might be offset by an increased likelihood of heat waves, drought, severe thunderstorms and tornadoes.

Several farming practices and technologies are available that can reduce greenhouse gas emissions and prevent climate change by enhancing carbon storage in soils, preserving existing soil carbon; and reducing carbon dioxide, methane and nitrous oxide emissions.

Innovative farming practices such as conservation tillage, organic production, improved cropping systems, land restoration, land use change and irrigation and water management, are ways that farmers can address climate change. Good management practices have multiple benefits that may also enhance profitability, improve farm energy efficiency and boost air and soil quality.

Carbon Credits

Emissions from agriculture contribute approximately 14% of global Greenhouse Gas (GHG) emissions and are expected to rise by 38% over the period 1990-2020.

Climate-smart agriculture (CSA) can help reduce emissions by sequestering carbon in trees and soils, and by reducing emissions from other sources such as land degradation, livestock and inefficient fertiliser use. At the same time CSA can help agriculture adapt to the impacts of climate change and increase productivity (and thus food security, farmers' incomes and agribusiness profits). CSA represents a 'triple win' situation: mitigation, adaptation and productivity. Thus, it can enhance agriculture sustainability, increase productivity, and resilience (adaptation), reduce/remove GHGs (mitigation), and enhances achievement of national food security and development goals.

Agriculture and Forest

Agricultural production is a major emitter of GHGs, currently accounting for 18% of total GHG emissions in India (INCCA, 2010). In 2019, the total livestock population is 535.78 million; cattle (192.90 million) is the largest animal group in the country followed by goats (148.88 million), buffaloes (109.85 million), sheep (74.26 million) and pigs (9.06 million). All other animals taken together contribute just 0.23 per cent of the total livestock population in the country. In 2019, the total livestock population registered a growth of 4.6 per cent over the last census in 2012 (512.06 million). The total population was 529.70 million at the time of 18th census in 2007.The cattle population has grown marginally by 0.83 per cent, and the buffalo population by 1.06 per cent. The populations of sheep (14.13%), goat (10.14%) and mithun

(26.66%) have risen significantly, underlining the preference of farmers for keeping milch animals. While the overall cattle population has increased by 0.8 per cent between 2012-19, the population of indigenous cattle has come down by 6 per cent — from 151 million to 142.11 million. However, this pace of decline is much slower than the 9 per cent decline between 2007 and 2012. In contrast, the population of the total exotic/crossbred cattle has increased by almost 27 per cent to 50.42 million in 2019.

Of this, 74% is due to methane produced from livestock, largely cows and buffalo, and rice cultivation. The remaining 26% comes from nitrous oxide emitted from fertilisers. Nearly two-thirds of the population relies on farming as the source of livelihood. India has 15-20% of the global cattle population, with around 996.36 million head cows and buffalo in 2018. It produces 19% of the world's milk. The country's share in world milk production stands at 17 per cent. India (66 million tonnes) and Pakistan (23.6 million tonnes) produce more than 90% of the total volume of buffalo milk in the world (as per the year 2012). India is also the second-largest producer of cow milk (54 million metric tons) followed by the USA

India will also need to substantially increase food grain production to feed its growing population, its climate pledge says. But droughts and floods are frequent, and the sector is already facing a high degree of climate variability.

India's forest cover has increased in recent years. As a matter of fact, India accounted for 7% of the global net increase in leaf area from 2000-2017. Its long-term goal is to bring 33% of its area under forest cover – some 109m hectares – up from 24% in 2013 (79m hectares). Another goal is to absorb 2,500-3,000 Mt CO_2 by 2030 through additional forest and tree cover. According to CAT, over half of this target could be achieved by the Green India Mission, launched in 2014, which aims to expand tree cover by 5m hectares and increase the quality of another 5m hectares of existing cover in 10 years. India's government also offers incentives for state action to increase forest cover by linking it to funding allocations.

The country could see $1.2tn of "lost GDP" in total, plus lower living standards for nearly half its population by 2050, compared to a scenario with no climate change, according to a 2018.

The Himalaya Mountains, which form the most important concentration of snow outside the poles, are one particularly vulnerable part of India and the wider South Asia region. The glaciers are a critical water source for 250 million people who live in the region. A further 1.65 billion people in India and seven other countries rely on the major rivers that flow from it.

Rising temperatures will melt at least a third of the region's glaciers by 2100, even if average global temperature rises are limited to 1.5°C. India is also vulnerable to increases in vector-borne diseases, such as malaria and dengue, which could both see increases due to climate change.

Heat waves are also expected to be an issue, with 600 million people currently in locations that could become moderate or severe hotspots by 2050, according to the World Bank. India's largely agricultural workforce was already significantly impacted by extreme heat in 2017, according to the Lancet Countdown on Health and Climate Change. India could see significant sea level rise, affecting the river water systems that hundreds of millions rely on for food. Its 1,238 islands are also at risk.

Impact of Climate Change on Agriculture

- Increased frequency and intensity of extreme climate events such as heat waves, droughts and floods, leading to loss of agricultural infrastructure and livelihoods.
- Decrease in freshwater resources, leading to water scarcity in arable areas.

- Water and food hygiene and sanitation problems.
- Changes in water flows impacting inland fisheries and aquaculture.
- Temperature increase and water scarcity affecting plant and animal physiology and productivity.
- Beneficial effects on crop production through carbon dioxide "fertilization".
- Detrimental effects of elevated tropospheric ozone on crop yields.
- Changes in plant, livestock and fish diseases and in pest species.
- Damage to forestry, livestock, fisheries and aquaculture.
- Acidification of the oceans, with the extinction of fish species.

The majority of Indian farming households depend upon the summer monsoon to irrigate their crops from June through September. Advance information about the likely character of the upcoming monsoon season could be very helpful both to farmers and policy makers. Farmers need information relevant to their specific region, and about specific climate variables affect crop management and agriculture policy decisions. In addition to seasonal and sub-seasonal rainfall forecasts, analysis of past climate patterns, such as the timing of monsoon onset and dry or wet spells, at appropriate scales can help decision makers better understand the climate risks they face.

Indian Agriculture: Scenario, the Emerging Impact and Vulnerability Assessment Climate Change in Agriculture Sector

Agriculture production is directly dependent on the general climatic conditions of a region and local weather. Possible changes in temperature, precipitation and CO_2 concentration significantly impact crop growth. The unpredictable variation in weather conditions and erratic occurrence of natural calamities on account of the phenomenon of climate change also impact agriculture. The overall impact of climate change on worldwide food production is, however, considered to be low to moderate, given the successful adaptation and adequate irrigation. Global agricultural production can increase, in fact, to the extent of doubling the CO_2 fertilization effect. Agriculture will also be naturally affected by the impact of climate change on water resources.

The Emerging Scenario

Climate has always played a significant role in the economic development of India. The country depends on climate-sensitive sectors like agriculture and forestry for livelihoods. The implications of climate change, therefore, assume significance to the countries' economies. Climate change can lead to geographical limits to agriculture, and changes in crop yields. Foodgrain production, which quadrupled during the post-independence era and is expected to continue to grow further, can adversely, in addition to impacting on food security and employment. Climate change can also increase the prevalence of pests.

Emerging Adaptations in Agriculture

Emerging adaptations in agriculture to climate change are diverse. It is categorized individual adaptation practices collected for this review into either of three categories: "incremental", "intermediate" or "transformational" for descriptive purposes. Incremental adaptation indicates adjustments of management to keep operating existing systems and it predominantly relies on existing technologies. Switching management practice from a conventional one to another which has already operated in another region of the world (warmer regions in particular) is an example of incremental adaptation. In contrast, transformational adaptation requires a change in the fundamental attributes of human and natural systems. Conventional management practices would be largely replaced with distinct new ones (e.g., switching crop types) when transformational adaptation occurs. Moving agricultural areas to other areas is another example of transformational adaptation. Intermediate adaptation falls within the middle of these two. Importantly, costs for adaptation (including transaction costs, opportunity costs and others; are expected to be substantial for transformational adaptation compared to those for incremental and intermediate adaptations.

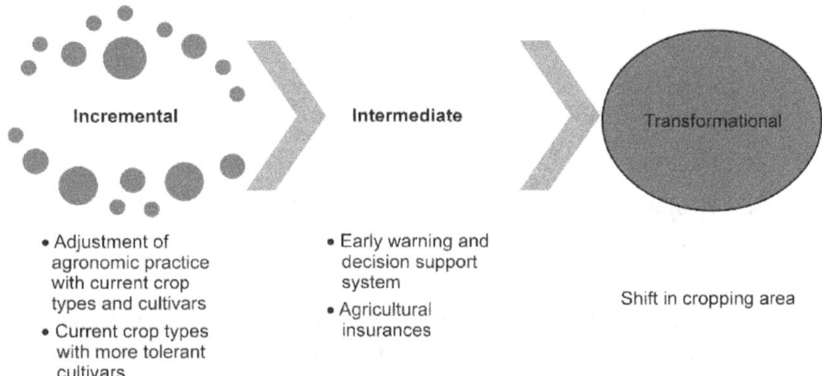

Figure 2.1 Transformational adoptions in Agriculture

During this 21ˢᵗ century, India will experience warming above the global level. It will also begin to experience more seasonal variations in temperature. The longevity of heat waves across India has increased in recent years with warmer night temperatures and hotter days, and the trend is expected to continue. The average temperature change is predicted to be 2.33°C-4.78°C, with a doubling in CO_2 concentrations. India's average temperature rose by around 0.7 degrees Celsius between 1901 and 2018. It is projected to rise further by approximately 4.4°C by the end of this century, according to the Indian Institute of Tropical Meteorology (IITM), Pune. These heat waves will lead to increased variability in summer monsoon precipitation, which will result in drastic effects on the agriculture sector in India. Models predict a gradual rise in carbon (CO_2) concentration and temperature across the globe.

Future Projections

The Inter-Governmental Panel for Climate Change (IPCC) projected a global average temperature rise of 4.2°C under the BAU (Business-As-Usual) emissions scenario (A1B) towards the end of the 21st century, while new studies project warming of more than 6°C under the current BAU emissions scenario over the same period.

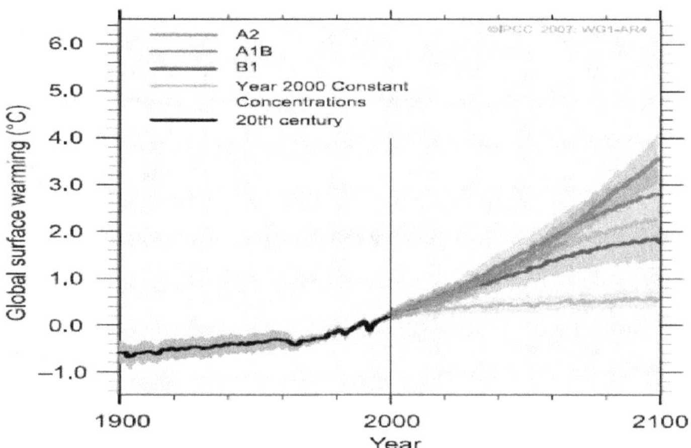

Figure 2.2 Projected Scenario of Climate Change

Source: Environment Statistics, 2019

Climate change depends on the balance of natural eco-systems (i.e., forests, river basins, sea level) and socio-economic systems (i.e., agriculture, fisheries, irrigation and power projects). Generally speaking, one finds that there is a limited allocation of resources, technology and finances, to the agriculture and allied sectors in developing countries such as India. Developing countries have very little capacity to develop and adopt strategies to cope with the challenges posed by climate change. It is also widely accepted that the poorest are disproportionately vulnerable to climate change and the least able to adapt. Therefore, research on the impact of climate change and vulnerability on agriculture needs to be accorded high priority.

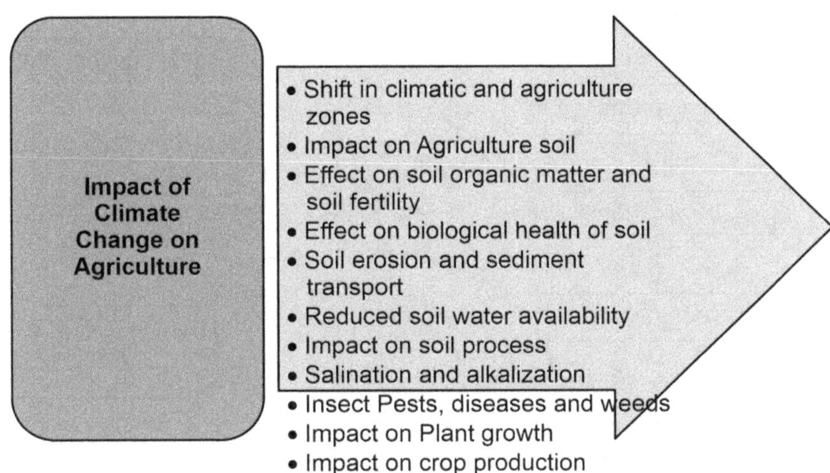

Figure 2.3 Impact of Climate Change on Agriculture

Indian agriculture now faces the dual challenge of feeding a billion people in changing climatic and economic scenarios. It has been projected that under the scenario of a 2.5°C to 4.9°C temperature rise, rice yields will drop by 32%-40% and wheat yields by 41%-52%. This would cause GDP to fall by 1.8%-3.4%. Agricultural productivity is sensitive in two broad classes of climate-induced effects (a) direct effects from changes in temperature, precipitation, or carbon dioxide concentrations and (b) indirect effects through changes in soil moisture and the distribution and frequency of infestation by pests and diseases. The impact assessment can be assessed in terms of three major factors i.e., environmental, biophysical and socioeconomic factors.

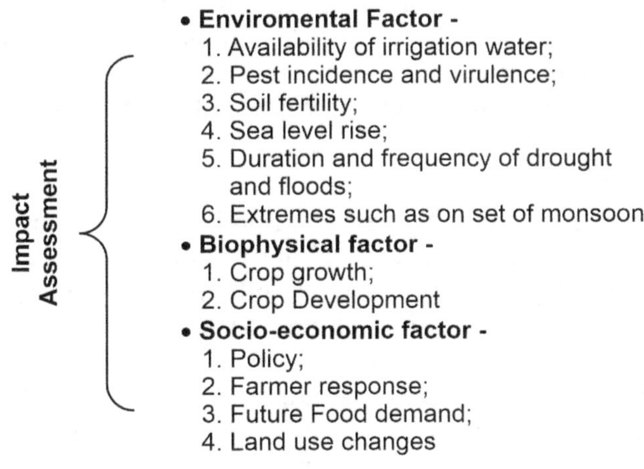

Figure 2.4 Impact Assessment of Agriculture

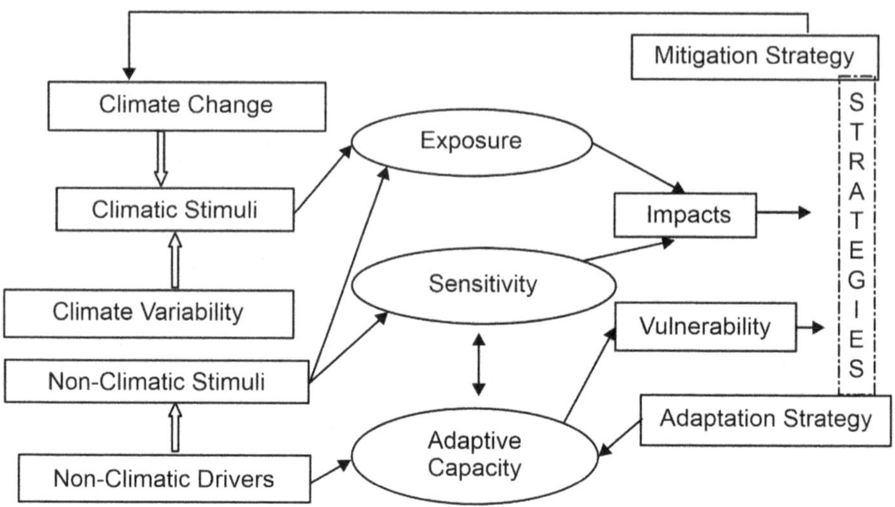

Figure 2.5 Assessment Framework: Impact and Vulnerability

Vulnerability

Vulnerability is the degree to which a system is susceptible to or unable to cope with adverse effects of climate change including variability and extremes. It is also seen as a function of the character, magnitude and rate of climate change and the degree to which a system is exposed, along with its sensitivity and adaptive capacity. The vulnerability of agricultural production to climate change depends not only on the physiological response of the affected plant but also on the ability of the affected socio-economic systems of production to cope with changes in yield as well as with changes in the frequency of droughts, floods, or cyclones, etc. The effect of climate change depends not only on the magnitude of climate stimuli or their effects but also on the sensitivity and capacity of the affected system to cope with or adapt to such stress is illustrated by the assessment framework of impact and vulnerability assessment.

Disaster Management

India is vulnerable, in varying degrees, to a large number of natural as well as man-made disasters. 58.6 per cent of the landmass is prone to earthquakes of moderate to very high intensity; over 40 million hectares (12 per cent of land) is prone to floods and river erosion; of the 7,516 km long coastline, close to 5,700 km is prone to cyclones and tsunamis; 68 per cent of the cultivable area is vulnerable to drought and hilly areas are at risk from landslides and avalanches. Vulnerability to disasters/emergencies of Chemical, Biological, Radiological and Nuclear (CBRN) origin also exists. Heightened vulnerabilities to disaster risks can be related to expanding population, urbanisation and industrialisation, development within high-risk zones, environmental degradation and climate change.

In the context of human vulnerability to disasters, the economically and socially weaker segments of the population are the ones that are most seriously affected. Within the vulnerable groups, elderly persons, women, children—especially women are rendered destitute, children orphaned on account of disasters and differently-abled persons are exposed to higher risks.

Vulnerability the Key Driver

India is a major disaster-prone country where about 85 per cent of the land is prone to some disaster or other. Physical, geographic and climatic conditions is among the world's most disaster-prone countries with almost two-thirds of the land area being vulnerable to single or multiple hazards. Out of 36 States/UTs some are disaster prone areas, and are vulnerable to windstorms formed in the Bay of Bengal and the Arabian Sea and earthquakes caused by active crustal movement in the Himalayan Mountains. Almost 58 per cent of the land is prone to earthquakes of moderate to very high intensity (high seismic zone III-V). India is also sensitive to floods and river erosions brought by monsoons and 68 per cent of the cultivable area is vulnerable to drought in the country's arid and semi-arid areas and the hilly regions are at risk from landslides and avalanches.

National Action Plan on Climate Change and Other Policies particularly Agriculture

The action plan outlines a number of steps to simultaneously advance India's development and climate change-related objectives. The Prime Minister's Advisory Council on Climate Change has outlined a National Action Plan on Climate Change (NAPCC) which was released by the Prime Minister in June 2008 and encompasses a range of measures. It focuses on eight missions, which are as follows:

1. National Solar Mission
2. National Mission for Enhanced Energy Efficiency
3. National Mission on Sustainable Habitat
4. National Water Mission
5. National Mission for Sustaining the Himalayan Ecosystem
6. Green India Mission
7. National Mission for Sustainable Agriculture
8. National Mission on Strategic Knowledge for Climate Change

National Mission for Sustainable Agriculture is one of the eight missions in the NAPCC which aims to support climate adaptation in agriculture through the development of climate-resilient crops, expansion of weather insurance mechanisms and agricultural practices. The mission focuses on four areas that are relevant to the endeavours of India's agricultural sector to adapt to climate change:

- Dry land agriculture.
- Risk management.
- Access to information.
- Use of technology.

The other policy is National Agriculture Policy, 2000 aims to attain over the next two decades a growth rate in excess of 4% per annum in the agriculture sector. The policy explicitly recognizes that agricultural growth should cater to domestic markets and maximize benefits from exports of agricultural products in the face of the challenges arising from economic liberalization and globalization.

Adaptation Strategies

Adaptation is to prepare communities, regions, countries and societies for the consequences of climate change. It is the process of adjustment by human and natural systems to natural and non-natural climate stimuli or their impacts. Its aim is to reduce the farmers' vulnerability and improve their adaptive capacity.

Table 2.1 Broad strategies and options of the agriculture sector.

Agricultural Research	• Agricultural research for development is fundamental to climate change adaptation. • A trans-disciplinary research approach with strong cooperation among national and international researchers is essential to find options for adaptation to climate change.
Agricultural Extension	• Agriculture Extension which requires • Education and innovation transfer.
Agricultural Insurance	• The increasing frequency of extreme weather calamities is causing financial losses on a large scale that traditional risk management strategies and social safety nets can no longer cope. • Agricultural insurances can supplement but, however, cannot replace comprehensive risk management strategies. • Insurance products that are suited for agriculture in the context of rural development land risk mitigation must be created.

Table 2.1 *Contd...*

Community-based adaptation to climate change	• Community-based initiatives assess the vulnerability of communities and help take measures to strengthen their resilience.
Agricultural Water Management	• Climate change influences the spatial and temporal availability of water, which in turn impacts food and fodder production. • Options in agricultural water management range from increasing the supply to managing the demand site.

Sustainable Development towards the Climate-Resilient Pathways

Sustainable development within the context of climate change calls for a new approach to development that considers the complex interactions between climate and society. Climate-resilient pathways recognize that impacts are certain because climate change can no longer be avoided. A variety of complex issues arise in the process of assessing climate-resilient pathways in a variety of regions at a variety of scales with sustainable development as the ultimate aim, mitigation as the way to keep climate change impacts moderate rather than extreme and adaptation as a response strategy to cope with impacts that cannot be avoided.

Pathways for sustainable development can be made more climate change resilient by risk management and vulnerability reduction strategies that include (a) Mitigation -Reducing the net rate of growth of Green House Gas (GHG) emissions and stabilizing or reducing their concentrations in the atmosphere and (b) Adaptation- improving capacities to cope with climate changes without disruptions of systems that we value. The linkage between adaptation and mitigation is very clear in the agricultural sector. Agriculture plays a dual role in climate change. It is not only a sector that will be severely affected but is also a significant contributor to global GHG emissions - the agriculture sector directly contributes to 14% of total GHG emissions.

Four basic mechanisms have been recognized for GHG mitigation in agriculture:

- Reducing methane and nitrous oxide emissions from agricultural production.

- Producing different forms of biomass for energy use as a substitute for fossil energy sources.

- Supporting forests through reforestation, afforestation and agroforestry.

- Storing carbon by increasing soil carbon content.

2.2 Disaster Management - Building a Sustainable Future

The outlook for the management of natural and man-made disasters in India in the years ahead, presents many clear challenges. While there are predictable natural disasters such as floods, drought and cyclones, there are also events such as earthquakes that cannot be predicted. In the long run, the challenges are even likely to become more significant. The impacts of global warming and climate change pose serious threats, to the world economy in general and to the agriculture sector in particular.

What these patterns show us is that we must put concerted attention on looking beyond response. The patterns of the last few years must teach us to expect the unexpected. What can we do differently to help reduce their impact, not just wait to respond once a disaster strikes?

In parallel, there is a need to look at risks and vulnerabilities that aren't yet technically 'disasters'. Considering the experiences of the years gone by, there will be impacts that continue to magnify under the radar – access to rapidly depleting water sources and the growing heat stress being two of the most prominent.

The manner in which these disasters are manifesting themselves and the resulting impact on communities is highly complex. For example, a single mega-disaster can wipe out hard-won development gains, as happened with the 2015 Nepal earthquake. Recurrent small-scale stresses keep the most marginalised families in a cycle of poverty. Climate change impacts are only magnifying these vulnerabilities. For at-risk communities and affected families, this interplay between dealing with poverty, climate stresses and natural hazards does not have clear distinctions. Each reinforces the other. This inter-connectedness is also already being recognised globally in international frameworks such as the Sustainable Development Goals and Paris Agreement.

The Rising Trend of Extreme Weather Events

Extreme weather events are on the rise in India, and we must plan and prepare for erratic weather patterns that are the new 'normal'. Almost every month of 2018 had one or the other 'unprecedented' weather events. We had hailstorms, unseasonal rainfall, strong thunderstorms and lightning, floods and droughts, long dry spells, cyclones, and both our monsoons — southwest and northeast — were below normal. Also, the average temperature over the country was 'significantly above normal', making 2018 the sixth warmest year on record since 1901. Thus, it would be no exaggeration to term 2018 as a year of multiple disasters; both in terms of 'visible ones' such as the catastrophic floods in Kerala in August, and 'silent ones' like an unprecedented monsoon rainfall deficit of more than 20% in the northeast region of the country.

Rising Temperature

Last year during 2019 started on a 'dry' note with very low winter precipitation in north India. There were reports of acute drought in Kashmir due to "a record-breaking long dry spell". Meanwhile, the spring came early in the Himalayas as Rhododendron arboretum, a small evergreen tree, started flowering in January, two to three months in advance. This was blamed on an unusual warm spell and rise in temperature in the hills. Rising temperature is not a one-off thing. Meteorological data analysed by the India Meteorological Department (IMD) shows a trend of rising temperature in large parts of the country. For instance, the IMD has carried out a long-term assessment of climate change in the country between 1951 and 2010. The state-wise averaged annual mean maximum temperature time series has shown increasing trends over many states of India, particularly significant over Andaman and Nicobar, Andhra Pradesh, Arunachal Pradesh, Assam, Goa, Gujarat, Himachal Pradesh, Jharkhand, Karnataka, Kerala, Lakshadweep, Madhya Pradesh, Maharashtra, Manipur, Mizoram, Odisha, Rajasthan, Sikkim, Tamil Nadu and Uttarakhand. The highest increase in annual mean maximum temperatures was observed over Himachal Pradesh (+0.06°C/year) followed by Goa (+0.04°C/year), Manipur, Mizoram and Tamil Nadu (+0.03°C/year) each.

Similarly, states that averaged annual mean minimum temperatures have shown significantly increasing trends over Andhra Pradesh, Arunachal Pradesh, Assam, Bihar, Delhi, Gujarat, Haryana, Kerala, Lakshadweep, Manipur, Meghalaya, Rajasthan, Sikkim, Tamil Nadu and Tripura. The highest increase in annual mean minimum temperature was observed for Sikkim (+0.07°C/year) followed by Arunachal Pradesh, Bihar, Delhi, Gujarat, Manipur, Tamil Nadu and Tripura (+0.02°C/year) each.

Early in 2018, Maharashtra, Chhattisgarh, Madhya Pradesh, Telangana and Uttar Pradesh were hit by unseasonal rains and hailstorms that damaged standing rabi (winter) crops in over 4.76 lakh hectare area. Maharashtra is prone to hailstorms, as documented in a countrywide analysis of hailstorms between 1985 and 2015 by the IMD, Pune office and scientists at the Met department pointed out that "from 2013, there is prolonged persistence of hailstorms in the state, causing extensive damage". Hailstorms now affect larger area in the state over several days. There are concerns over the increasing size of hail, too.

There are studies that link an increase in hailstorm activity with the changes in the climate. According to a 2010 study, 'Climate change and hailstorm damage: Empirical evidence and implications for agriculture and insurance', hailstorm damage may increase in the future if global warming leads to further temperature increase. Our estimates show that by 2050, annual hailstorm damage to outdoor farming could increase by between 25% and 50%, with considerably larger impacts on greenhouse horticulture in summer by more than 200%".

Agro-Meteorological Challenges due to Extreme Weather Events

India has been repeatedly battered by extreme events like heavy rainfall causing extensive flooding, droughts, unseasonal rainfall, hailstorm, etc. The period between 2001–2015 witnessed the intensification of climate and weather extremes such as destructive flooding, severe droughts, heat waves, heavy rainfall and severe storms in different regions. The number of extreme events of very heavy rainfall has almost doubled in the country in the last 50 years. Among other extreme events, unseasonal rains and hailstorms are mostly observed during the pre-monsoon season from March to April in the country. In some years, it has occurred early during the end of February and late during mid-May also. The unseasonal rains and hailstorms have destroyed crops in millions of hectares of farmland in many states including Himachal Pradesh, Uttar Pradesh, Uttarakhand, Punjab, Haryana, Madhya Pradesh, Gujarat, Rajasthan, Maharashtra and Andhra Pradesh, causing huge losses to farmers. Also, hailstorm causes substantial damage to the standing crops as well as the horticultural crops within a very short time span.

Though occurrences of hailstorms are unavoidable, there is a need for its prediction followed by recovery, rescue and remedial measures. There are modern methods available to detect hail-producing thunderstorms using weather satellites and weather radar imagery. Also, severe weather warnings are issued by Indian Meteorological Department (IMD) for hailstorms when the hail reaches a damaging size, as it can cause serious damage to structures, crops and livestock. The accurate and timely information on extreme meteorological parameters has not only great potential for increasing output to the farmers, but also is useful for modification of crop environment, protection from frost, strong wind, and also irrigation scheduling leading to efficient water management and drought preparedness.

With recent advances in computational power and prediction modelling and technology, it is now possible to forecast the occurrence of extreme events and the nature of devastation that they may cause with a greater degree of accuracy and with a longer lead time. The World Meteorological Organization has launched the Global Framework for Climate Services to use climate information services to meet the challenges of the future particularly with reference to extreme events in five focus sectors including agriculture. IMD has also established Climate Research and Services as part of the national mechanism in this regard. The provision of need based climate information to farmers can support the management of agricultural resources (land, water and genetic resources). In India, a combination of traditional and more innovative technological approaches are being used to manage drought risk. Also, technological drought management (e.g., development and use of drought-tolerant cultivars, shifting cropping seasons in agriculture, and flood and drought control techniques in water management) is combined with model-based seasonal and annual to decadal forecasts. The model results are then translated into an early warning in order to take appropriate drought protection measures.

Thunderstorms and Lightning

After hailstorms, large parts of the country faced a series of extremely strong thunderstorms and dust storms last year. A thunderstorm, mostly a short-duration phenomenon that seldom lasts over two hours, is always accompanied by thunder and lightning, usually with strong gusts of wind, heavy rain, and sometimes with hail. India recorded a total of 3,620 thunderstorms in 2014 and 5,536 in 2015. The genesis of a thunderstorm is dependent on four broad factors – intense heating, moisture availability, instability in the atmosphere, and a trigger.

Last year, Down To Earth, a fortnightly magazine, documented how 44 intense thunderstorms struck over 16 states killing 423 people, which was 'unprecedented'. Thunderstorms were accompanied by lightning strikes that destroyed properties and killed people. Data from the National Crime Records Bureau shows the average number of people dying of lightning strikes every year between 2006 and 2015 was about 50% higher than the decade before. Scientists at the Indian Institute of Tropical Meteorology, Pune have examined satellite data between 1990 and 2013 and found a two to three percent increase in lightning strikes in the country.

Experts have attributed the 'above normal' thunder activity over Northwest India last year to the late western disturbances. As per the IMD, seven western disturbances passed over the region within two months (April and May 2018) – including three intense ones in the first half of May. This seems higher than usual as a 2017 study has documented an average of two to three western disturbances a month across Northwest India in the summer months.

Climate scientists say western disturbances used to be very active in the month of January over North India. But these seem to be getting pushed to the spring season. While there isn't enough past data to reach a definitive conclusion, this area of changing pattern of western disturbances needs more research as it directly impacts our weather. Apart from western disturbances, jet streams also impact our weather and may be linked to increasing storms. Jet streams, or river-like currents of air, circulate in the upper levels of the troposphere (30,000 feet). A jet stream exists because of the temperature difference between the Poles and the tropical regions. Since the Arctic is warming at double the rate of the rest of the world, this temperature difference is reducing, thereby affecting the jet streams. There are scientists who attribute the increasing intensity of storms to the rising heat, too. For instance, urban heat island is expected to have an impact on thunderstorms, as surface temperatures have fluctuated over Indian cities because of land-use and land-cover changes. This impacts the lower atmosphere and can influence the trigger mechanism of thunderstorms.

Monsoon: Redefining 'normal'

The southwest monsoon in 2018, our main monsoon season (June to September) when the entire country receives rainfall that supports agriculture, ended at a 'below normal' note. That too after the IMD's long-range forecast had predicted a 'normal' monsoon. All through the four months of the southwest monsoon, there were prolonged dry spells affecting Kharif (summer) crop sowing and plant growth, and an almost dry September. By the end of September, the northeast region of the country ended with a rainfall deficit of more than 20%. At an all India level, the southwest monsoon rainfall deficit was 9%, classified as 'below normal'. However, meteorologists point out that the definition of 'normal' monsoon itself is tricky and needs to be updated. 'Normal' rainfall is defined as 96%-104% of the long period average (LPA), with a model error of plus or minus 5%. The LPA is the weighted average of rainfall that India received in June-September from 1951 to 2000 and is pegged at 89 cm. The IMD issues two long-range forecasts for the southwest monsoon. The first stage forecast is issued in April and the second stage in early June. The second stage forecast covers the four homogenous regions of the country - Central India, Peninsular India, Eastern India and North-western India - and provides monthly monsoon rainfall forecast for the months from June to September.

An analysis of the rainfall data of the past few years shows that several subdivisions received deficient rainfall even in normal monsoon years. For instance, the country received 95% rainfall of the LPA in 2017, but Vidarbha had 23% deficient rainfall. The East Madhya Pradesh subdivision reported 24% less rainfall than normal, East Uttar Pradesh 28% shortfall and West

Uttar Pradesh 30% shortfall. The Haryana, Delhi and Chandigarh subdivision and the Punjab subdivision also had deficient rainfall. On the other hand, the subdivisions of West Rajasthan and Saurashtra and Kutch received excess rainfall.

Droughts and Floods

A 'below normal' monsoon rainfall during 2018 triggered droughts in several states including Bihar, Jharkhand, Karnataka, Maharashtra and parts of Andhra Pradesh, Gujarat and Rajasthan. Maharashtra has sought drought-relief of ₹ 7,000 crore from the Centre. Karnataka and Jharkhand, too, have asked for ₹ 2,434 crore and ₹ 819 crore drought relief, respectively. What makes the matter worse is that the last year ended with a 'below normal' northeast monsoon, too. Unlike the southwest monsoon, the northeast monsoon (October to December) brings rainfall to some meteorological subdivisions in the south peninsula. In its January 16, 2019 'Statement on Climate in India during 2018', the IMD pointed out that rainfall during the northeast monsoon, between October and December, over the country had been 'substantially below normal' – only 56% of the long-term average. And this, it said, was the sixth lowest since 1901.

While almost half the country is facing drought, there were states that also faced floods last year. The situation is grim in the northeast region where states such as Assam, Arunachal Pradesh and Nagaland, which had faced floods, are now dealing with drought conditions.

Cyclones

While several states were declared drought-affected, there were over 14 depressions and four cyclones last year — Daye (September), Titli (October), Gaja (November) and Phethai (December). Cyclone-battered Tamil Nadu (Gaja) and Andhra Pradesh (Titli) sought aid of ₹ 15,000 crore and ₹ 1,200 crores, respectively. There are several research studies that link changing climate to an increasing frequency of intense tropical cyclones in the north Indian Ocean. Data of 122 years of tropical cyclone frequency over the north Indian Ocean from 1877 to 1998 shows "there is indeed a trend in the enhanced cyclogenesis during November and May". There has been a two-fold increase in the tropical cyclone frequency over the Bay of Bengal, a 17% increase in the intensification rate of cyclonic disturbances to the cyclone stage and a 25% increase in severe cyclone stage over the north Indian Ocean during November in the past 122 years, the study notes.

Analysis by Down To Earth suggests that extreme weather events have increased from just one from1900-1910 to 61 during 1971-80. And that number almost tripled to 162 during 2001-2010. While no single extreme weather event can be attributed to climate change, it is also true that the increased frequency and intensity of these extreme events are due to human induced climate change. Extreme rainfall, extreme heat, extreme thunderstorms, extreme cold waves, extreme tropical storms are the new 'normal'. And, we need adaptation strategies and disaster response plans to face exigencies arising out of this new 'normal'. To address the increase in extreme weather events, the earth sciences ministry is developing improved models for short-range weather forecasts, and a prediction system for thunderstorms and lightning. It is further improving the prediction of tropical cyclones by developing a coupled (ocean-atmosphere) model. Also, 10 Doppler Weather Radars are being installed in three hilly states of northwest India. Another 11 radars are expected to be installed over the plains, including one in Mumbai by early 2020.

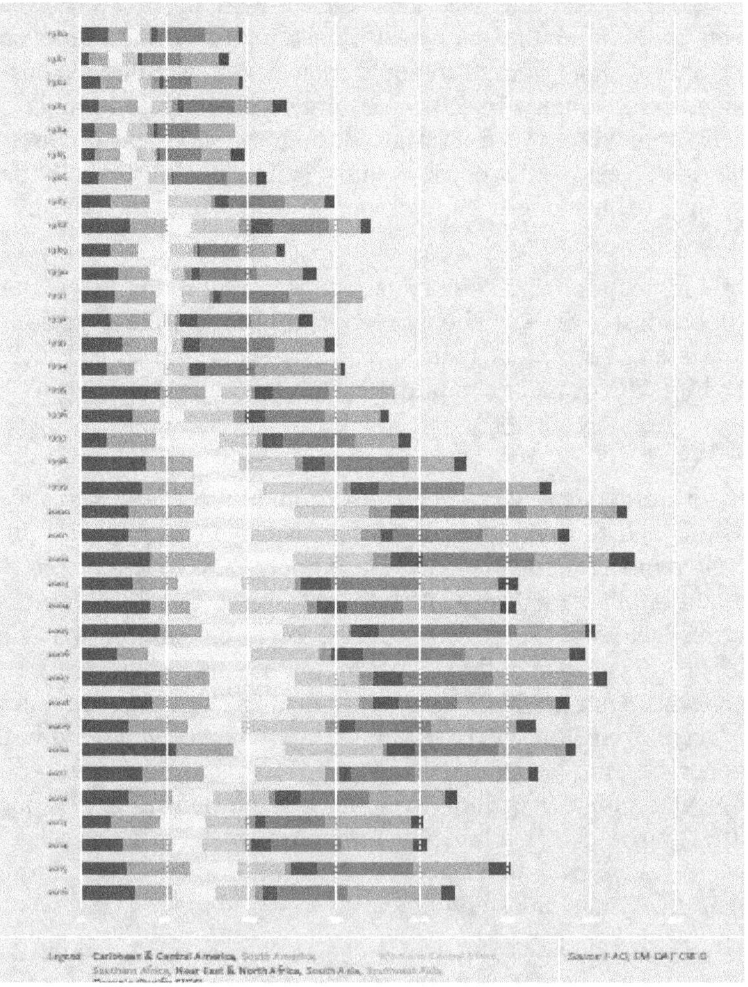

Figure 2.6 Occurrence of natural hazard-induced disasters in developing countries, 1980 – 2016

Source: SAO

Trends in hazard patterns, vulnerabilities and capacities strongly indicate a complex road ahead, wherein risk reduction efforts will need to be very substantially strengthened with the use of improved governance, people's participation and innovative processes and technologies.

Natural Disasters and Food Chain Crises

Since 1980, natural disasters have hit every continent and region of the world with growing frequency and intensity. The number of recorded natural disasters, along with their associated impact on livelihoods and economies at both local and national levels is increasing significantly. These include geophysical disasters, climate and weather-related disasters as well as outbreaks of animal and plant pests and diseases (the biological disasters behind food chain emergencies). On a global level, the economic loss associated with such disasters now averages between USD 250 billion to USD 300 billion every year. In developing countries, an average of 260 natural disasters occurred per year between 2005 and 2016, taking the lives of 54 000 people on average each year, affecting over 97 million others and costing an average of USD 27 billion in economic loss annually (EM-DAT CRED).

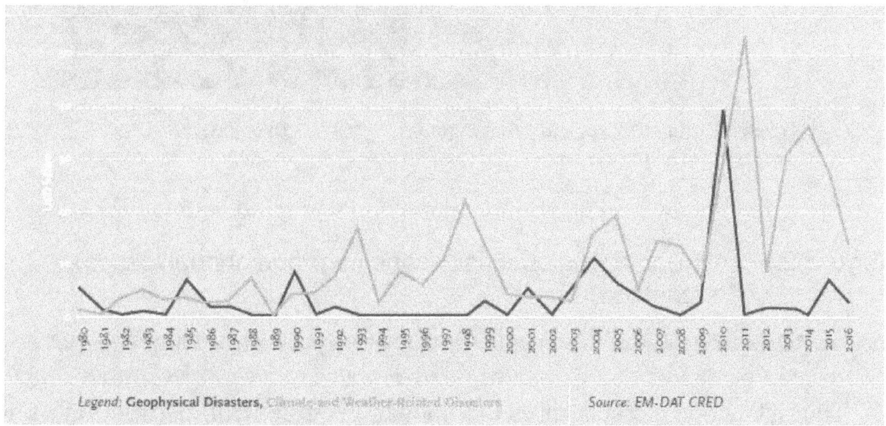

Figure 2.7 Economic loss from disasters in developing countries: geophysical vs. climate- and weather-related disasters, 1980–2016

Source: EM-DAT CRED

There were 260 natural disasters (both climate- and weather-related as well as geophysical and biological) per year in developing countries between 2005 and 2016 – an 11 percent increase on the 1993–2004 period when the average was 235 per year, and a more than a two-fold increase on 1981–1992 when they averaged 122 per year. While the economic impact of geophysical disasters (earthquakes, tsunamis, volcanic eruptions and mass movements) has remained fairly stable over the past decades, annual economic loss from climate and weather-related events has been consistently growing, in line with the increasingly frequent occurrence of the latter. Though the damage and loss have not yet been calculated, 2017 – the most violent hurricane season on record – will certainly confirm this trend.

New Regional Risk Scope

The key takeaway is that the economic losses due to disasters are larger than those previously estimated with most of this additional loss linked to the impact of slow-onset disasters in the agricultural sector. The multi-hazard average annual loss for the region is $675 billion, of which $405 billion, or 60 per cent, is drought-related agricultural losses, particularly in rural economies.

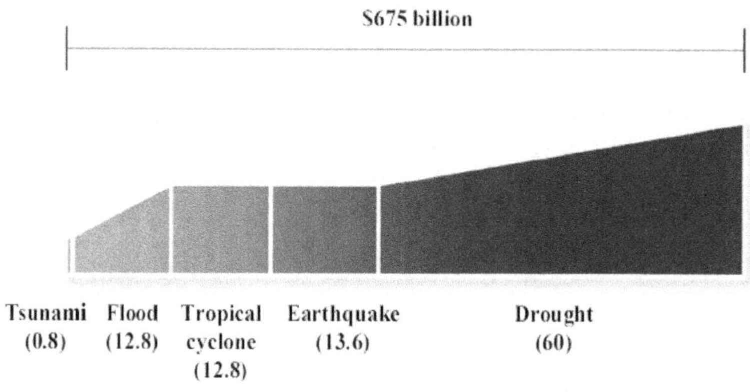

Figure 2.8 Asia-Pacific regional risk scope (average annual losses) (Percentage)

Source: Asia-Pacific Disaster Report 2019 (United Nations publication, Sales No. E.19.II.F.12).

The risk scope also captures the uneven geographical distribution of average annual loss for individual hazard types. Of the region's total earthquake related average annual loss, 64 per cent is in Japan and 14 per cent is in China. For tropical cyclones, approximately half the damage is in Japan, followed by 16 per cent in the Republic of Korea, 14 per cent in the Philippines and 13 per cent in China. For flooding, China represents 28 per cent of the average annual loss, and India 13 per cent, followed by the Russian Federation at 9 per cent and Australia at 7 per cent. For tsunamis, almost all damage is found in Japan.

Countries can also be ranked in terms of multi-hazard average annual loss. On this basis, the five countries at greatest risk of rapid-onset disasters are Japan, China, the Republic of Korea, India and the Philippines. However, the picture changes when slow-onset disasters are added. The new order is led by China, followed by Japan, India, Indonesia and the Republic of Korea.

The inclusion of slow-onset disasters, therefore, substantially changes the understanding of the geography of risk in the region, as populous countries move up the ranking.

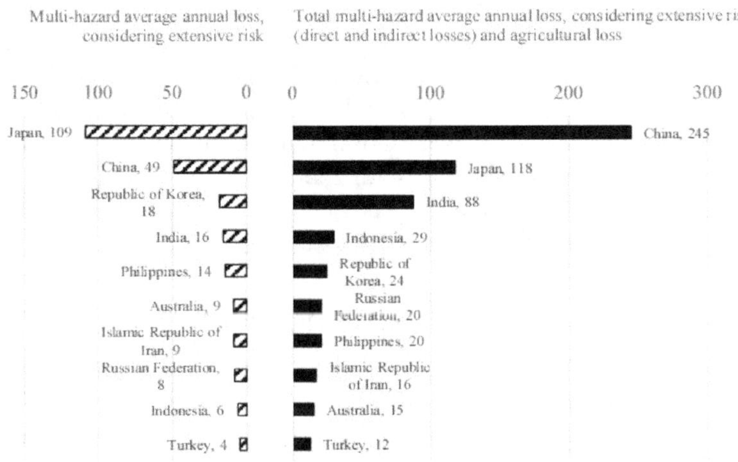

Figure 2.9 Risk Scape in numbers: average annual loss (Billions of United States dollars)

Source: Asia-Pacific Disaster Report 2019.

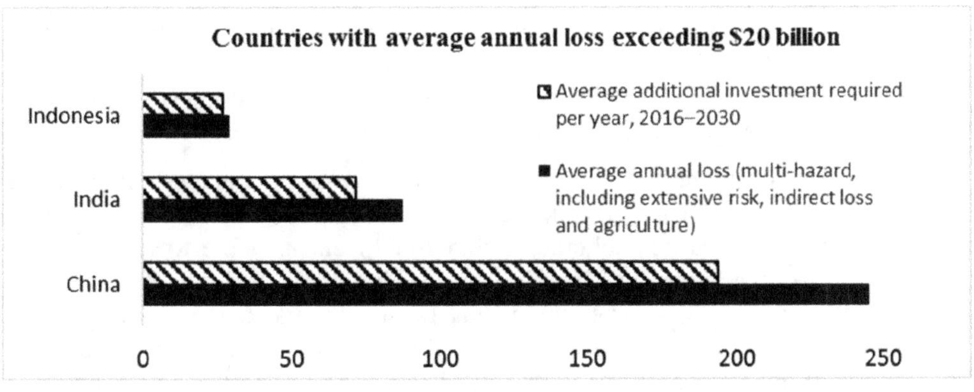

Figure 2.10 Countries with average annual loss exceeding $ 20 Billion
Source: Asia-Pacific Disaster Report 2019.

2.3   Management of the Agriculture sector in COVID-19 lockdown period

Coronavirus disease (COVID-19) is an infectious disease caused by a newly discovered coronavirus. Most people infected with the COVID-19 virus will experience mild to moderate respiratory illness and recover without any special treatment. Older people, and those with underlying medical problems like cardiovascular disease, diabetes, chronic respiratory (i.e., lungs) disease, and cancer are more likely to develop serious illness. The COVID-19 virus spreads primarily through droplets of saliva or discharge from the nose when an infected person coughs or sneezes. The Coronavirus pandemic is a public health emergency with grave implications for the population around the world. India, as part of the global community has been adversely impacted by the health crisis especially in the poor and marginalised.

COVID-19–Impact on Indian Agriculture

The health crisis around COVID-19 has affected all walks of life. Protecting the lives of people suffered from the disease as well as frontline health responders has been the priority of the nation. During the challenging time, how does Indian Agriculture respond to the crisis and how do government measures affect 90.2 million agricultural households across India. Thereafter, impact the economy of a country in a developing world. Immediately the nation-wide lockdown was announced, and the Indian Finance Minister declared a package, to protect the vulnerable sections (including farmers) from any adverse impact of the Coronavirus Pandemic. Similarly, the sale of dairy products, fish, poultry, etc., has been hit during the lockdown period as the uptake by the organised industry players affected due to shortage of workforce and transportation issues. As weather has been very erratic over the past few months in many parts of India, harvested produce protection must be a risk.

The Government raised the wage rates for workers engaged under Mahatma Gandhi National Rural Employment Guarantee Act (NREGA), the world's largest wage guarantee scheme. Under the special scheme to take care of the vulnerable population, Pradhan Mantri Garib Kalyan Yojana (Prime Minister's scheme for welfare of the poor), has been announced. Likewise, the Indian Council of Agriculture Research (ICAR) has issued state-wise guidelines for farmers to be followed during the lockdown period. The Reserve Bank of India (RBI) has announced specific measures that address the "burden of debt servicing" due to the COVID-19 pandemic. Agricultural term and crop loans have been granted a moratorium of three months (till May 31, 2020) by banking institutions with 3 percent concession on the interest rate of crop loans up to ₹ 3,00,000 for borrowers with good repayment behaviour.

Aatma Nirbhar Bharat Abhiyaan

The GoI announced a special economic package of ₹ 20 lakh crore (equivalent to 10% of India's GDP) with the aim of making the country independent against the tough competition in the global supply chain and to help in empowering the poor, labourers, migrants who have been adversely affected by COVID-19. Subsequent announcement of the detailed measures spelt out under the economic package sector is given below:

Concessional Credit Boost to Farmers: Farmers will be provided institutional credit facilities at concessional rates through Kisan Credit Cards (KCCs). This scheme will cover 2.5 crore farmers with concessional credit worth two lakh crore rupees.

Agri Infrastructure Fund: A fund of one lakh crore rupees will be created for development of agriculture infrastructure projects at farm-gate and aggregation points (such as Cooperative Societies and Farmer Producer Organizations). Farm gate refers to the market where buyers can buy products directly from the farmers.

Emergency Working Capital for Farmers: An additional fund of ₹ 30,000 crore will be released as emergency working capital for farmers. This fund will be disbursed through National Bank for Agriculture and Rural Development (NABARD) to Rural Cooperative Banks (RCBs) and Regional Rural Banks (RRBs) for meeting their crop loans requirements. This fund is expected to benefit three crore small and marginal farmers. This is in addition to the financial support of ₹ 90,000 crore that will be provided by NABARD to RCBs and RRBs to meet the crop loan demand this year.

Support to Fishermen: The Pradhan Mantri Matsya Sampada Yojana (PMMSY) will be launched for integrated, sustainable, and inclusive development of marine and inland fisheries. Under this scheme, ₹ 11,000 crore will be spent on activities in marine, inland fisheries and aquaculture and ₹ 9,000 crore will be spent on developing infrastructure (such as fishing harbours, cold chain, and markets).

Animal Husbandry Infrastructure Development: An Animal Husbandry Infrastructure Development Fund of ₹ 15,000 crore will be set up, with the aim of supporting private investment in dairy processing, value addition, and cattle feed infrastructure. Incentives will be given for establishing plants for the export of niche dairy products.

Employment Push using CAMPA Funds: The government will approve plans worth ₹ 6,000 crore under the Compensatory Afforestation Management and Planning Authority (CAMPA) to facilitate job creation for tribal/Adivasis. 2 Funds under CAMPA will be used for: (i) afforestation and plantation works, including in urban areas, (ii) artificial regeneration, assisted natural regeneration, (iii) forest management, soil and moisture conservation works, (iv) forest protection, forest and wildlife-related infrastructure development, and wildlife protection and management. Note that the CAMPA funds are currently used for the protection of forest and wildlife management.

Amendments to the Essential Commodities Act: The Essential Commodities Act, 1955 empowers the central and state governments to control the production, supply and distribution of certain commodities to avoid artificial scarcity in the country. Commodities covered under the Act include edible oil and seeds, pulses, sugarcane and its products, and paddy. The Act will be amended to deregulate food items including cereals, edible oils, oilseeds, pulses, onions and potatoes. This is expected to allow better price realisation for farmers by attracting investments and enabling competition in the sector. The stock limit will be imposed under very exceptional circumstances such as national calamities and famines with a surge in prices. Further, no such stock limit will apply to processors or value chain participant, subject to their installed capacity, or to any exporter subject to the export demand.

Agriculture Marketing Reforms: A central law will be formulated to provide: (i) adequate choices to farmers to sell their produce at remunerative prices, (ii) barrier-free inter-state trade, and (iii) a framework for e-trading of agriculture produce. Currently, farmers are bound to sell their produce only to the licensees in Agricultural Produce Market Committees (APMCs). The proposed amendments seek to enable the free flow of agricultural produce and establish a smooth supply chain providing options of better price realisation to farmers.

Agriculture Produce Pricing and Quality Assurance: A facilitative legal framework will be created to enable farmers to engage with processors, aggregators, large retailers, and exporters in a fair and transparent manner. Risk mitigation for farmers, assured returns, and quality standardisation will form an integral part of the framework. This is aimed at enabling farmers to predict the price of crops at the time of sowing and will also increase private sector investment in the sector.

Focussed Policies Address the Core Challenges of Agriculture

The GoI in its Union Budget focused on boosting income and enhancing purchase power in urban, semi-urban India (new personal tax regime), and rural India (through a push for agriculture, irrigation, aquaculture, horticulture, etc.). Oriented towards doubling farmers' income by 2022. Significant initiatives announced for agriculture, irrigation and rural development are aimed at doubling farmer's income by 2022 – Fasal Bima Yojana (FBY), Solar pumps for farmers, promotion of warehouse, cold storage, agriculture credit target of ₹ 15 lakh crores for the financial year 2020-21, horticulture support, fishery exports, blue economy push, cattle fodder farms, milk processing, etc., are all measures, if implemented well, likely to uplift the earnings of people in rural India thereby benefiting consumer and retail companies especially in Fast Moving Consumer Goods (FMCG) sector.

The Union Budget of India for 2020-21 provides a comprehensive 16-point action plan for further giving a boost to the Government's goal of doubling the farmers' income by 2022. The 16-point plan focuses on the core and fundamentally critical challenges facing the agriculture sector like financial inclusion and access to formal channels of credit, logistical access for perishable farm produce through the support of the Indian railways' transportation, handholding to farmers at water stressed districts, improved storage facilities, better linkage of the market resulting in the creation of an additional market for farm outputs, reduction of post-harvest losses, sustainable farming, etc. The agriculture policies implemented well by the Central and State government would gradually contribute to the inclusive development of India and should give a boost to rural India and in turn increase India's GDP exponentially.

16-Point Action Plan for Doubling Farmer's Income by 2022

1. Push for uniform agriculture laws across the country by encouraging states to implement central Model Agriculture Laws.

2. Hand-holding farmers at 100 water-stressed districts by providing them with comprehensive measures.

3. Reduction of cost through the use of solar technology by expanding PM-KUSUM scheme to support 20 lakh farmers in setting up stand alone solar pumps and 15 lakh farmers in solarising their grid-connected pump sets. In addition, a new scheme is to be operationalized to set-up solar power generation units on fallow/barren lands.

4. Encourage balanced use of traditional organic and other innovative fertilizers to reduce excessive use of chemical fertilizers.

5. Exercise to map and geo-tag agri-warehouses, cold storage and other inventory storage, create warehousing and provide Viability Gap Funding through PPP mode where State to facilitate land for setting up efficient warehouses at block/ taluka level.

6. Proposal to start a Village Storage Scheme run by SHGs which will help in the reduction of logistics costs and post-harvesting losses.

7. Setup a Kisan Rail through the PPP mode for seamless transportation and faster access to a market of perishable goods inclusive of milk, meat and fish and inclusion of refrigerated coaches in Express and Freight trains.

8. Introduction of Krishi Udaan to boost agri-exports and ensure better connectivity within Indian agri-markets like North-east and tribal areas.

9. Focus on "One Product – One District" clustered approach in the horticulture sector for better marketing and ease in export.

10. Focus on Zero budget Natural farming and expansion of integrated farming systems in rainfed areas.

11. Integration of financing of negotiable warehousing receipts with e-NAM.

12. NABARD Refinancing Scheme to further expand with the Kisan Credit Card scheme which will help in improved credit facilities for farmers.

13. Measures to eliminate Foot and Mouth disease, Brucellosis in cattle and Peste des Petits Ruminants (PPR) in sheep by 2025 will improve the productivity of livestock. Milk processing capacity will be doubled by 2025.

14. Framework for Blue Economy, i.e., development, management and conservation of marine fishery resources, which has registered more than double growth in the past 5 years and has emerged as the largest group in agro-exports.

15. It is endeavoured to increase fishery exports and promote the growth of algae, seaweed and cage culture and raising of fish production in India through 3477 Sagar Mitras and 500 Fish Farmer Producer Organisations.

16. Further mobilisation of Self-Help Groups (SHGs) under Deen Dayal Antyodaya Yojana (DDAY) for the alleviation of poverty.

Strong implementation of the 16-point action plan by the Central and State Government for the Agriculture Sector would strongly address the root challenges of the Indian Agriculture Sector, thereby impacting 70% of India's households who primarily depend on agriculture for their livelihood. Ensure India's food security, very large impact on strengthening the value chain, i.e., farm to fork, thereby giving a direct boost to the promising Indian food processing sector. Strengthen the rural economy, consequently positively impacting all the other sectors of the Indian economy.

The ₹1 lakh crore useful to farmers for the development of agriculture infrastructure projects at farm-gate, aggregation points, establish primary agricultural cooperative societies, Farmer Producer Organisations (FPOs), agriculture entrepreneurs and start-ups. This will provide an impetus for development for farm-gate or aggregation point affordable and financially viable post-harvest management infrastructure, Minimum Support Price (MSP), crop insurance, Kisan Credit Cards, useful to fishermen and animal husbandry farmers, access to institutional credit at concessional interest rates, etc.

COVID-19 Impact on Indian Food Sector

Agriculture is the backbone of the country and part of the Government announced essential category; the impact was likely to be low on both primary agricultural production and usage of agri-inputs like seeds, pesticides and fertilisers. Migratory labour movement for harvesting paddy, wheat, pulses, etc., should have been allowed for the harvesting rabi season crops in different states. Insulating the rural food production are as in important for containing the macro impact of COVID-19 on the Indian food sector as well as the economy in general. Crop procurement and mandi operations were yet to be streamlined and this would result in low sowing in the upcoming crop season and also would impact the sale of agri inputs in the Kharif season.

Agriculture plays a vital role in India's economy with 54.6% of the total workforce is engaged in agricultural and allied sector activities and accounts for 17.1% of the country's Gross Value Added (GVA) for the year 2017-18 (at current prices). It has 1457.26 lakh farm holdings and more than 80% of them are small and marginal farmers. As per the Land Use Statistics 2014-15, the total geographical area of the country is 328.7 million hectares, of which 140.1 million hectares is the reported net sown area and 198.4 million hectares is the gross cropped area with a cropping intensity of 142%. The net area sown works out to 43% of the total geographical area. The net irrigated area is 68.4 million hectares. The agriculture and allied sector witnessed a growth of 2.7 per cent in 2018-19 at 2011-12 basic prices.

In such a situation India had to face the pandemic of Corona Virus. The first case of the 2019–20 coronavirus pandemic in India was reported on 30 January 2020, originating from China. The outbreak has been declared an epidemic in the states and union territories, where provisions of the Epidemic Diseases Act, 1897 and National Disaster Management Act, 2005 have been invoked, and educational institutions and many commercial establishments have been shut down. India has suspended all tourist visas, as a majority of the confirmed cases were linked to other countries.

On 22 March 2020, India observed a 14-hour voluntary public curfew. The government followed it up with lockdowns, where COVID cases had occurred as well as all major cities. Further, on 24 March, ordered a nationwide lockdown for 21 days, affecting the entire 1.3 billion population of India. It had huge effects on migrant labourers, informal workers, micro and small enterprises, farmers and the self-employed, who are left with no livelihood in the absence of transportation and access to markets.

The National Disaster Management Authority (NDMA) in order No. 1-29/2020-PP (PP-II) Dt: 24/03/2020 after satisfying that the country is threatened with the spread of COVID-19, which has been declared as a pandemic by the World Health Organization and has considered it necessary to take effective measures to prevent its spread across the country and that there is need for consistency in the application and implementation of various measures across the country while ensuring maintenance of essential services and supplies etc. Based on the above orders of NDMA, the Ministry of Home Affairs (MHA), GoI in order No.40-3/2020-DM (A) Dt: 24/03/2020 have issued orders for strict implementation of the above directions. GoI (MHA) further issued detailed instructions vide order No.40-3/2020-DM-I(A), Dt: 26/03/2020 exempting to lock down measures viz., ration shops, fruits and vegetables, seeds and pesticides and also fertilizer as Essential Commodities.

Agriculture Inputs

MHA, GoI, have exempted supply of Seeds, Fertilizers and Pesticides during the lockdown period.

1. **Seeds:** states have cleared this as an essential category and impact is likely to be low.

2. **Agro – Chemicals:** companies that depended on exports of finished goods for sale and imports of raw ingredients would be impacted.

3. **Fertilisers:** owing to existing inventories might be less impacted except for logistic import clearance in India.

Coronavirus attack created an unprecedented situation on the economy is no doubt devastation. No sector escaped with COVID-19 impact and its impact on agriculture is complex and varied across diverse segments that form the agricultural value chain. Even among the different segments, its impact varies widely among different regions and among producers and agricultural wage labourers. Some of the problems during the lockdown period are labour non-availability and inability to access markets for the produce due to issues in transportations as well as the operation of markets.

Some parts of State agriculture that have the luxury of deployment technology for harvests, like paddy and wheat are relatively more insulated since farmers regularly do not have to depend on a large number of manual labour. The increased use of mechanical harvesters for paddy helped in the present circumstances, though their inter-state movements have severely curtailed. However, commercial crops are drastically hit as they tend to be more dependent on migrant labour. Consequently, the shortage of migrant labour has resulted in a sharp increase in daily wages for harvesting crops. In many states, the rise is as high as 50 per cent, makes it unremunerative for producers since prices collapsed due to either lack of market access includes the stoppage of transportation and closure of borders. This is in contrast to areas where migrant labourers returned home from urban areas and led to a sharp decline in agriculture wages.

Agriculture producers are particularly hard hit with return on produce vary from one-third the usual or a complete loss. In a number of districts, inter-state trade in commercial crops or proximity to urban areas provide market access and better prices. These are often due to initiatives of individual farmers rather than direct state support. This is frequently the case of crops like onion, cotton, mango, flowers, vegetables and inland fisheries. The rise in labour costs and lack of access means the farmers are staring at huge losses and hence it allows crops to rot in the field, a better "stop-loss" mechanism.

The economic shock will be much more severe for India, for two reasons: 1. Pre COVID-19, the economy was already slow down, compounding existing problems of unemployment, low incomes, rural distress, malnutrition, and widespread inequality; 2. India's large informal sector is particularly vulnerable. Out of the national total 465 million workers, around 91 per cent (422 million) were informal workers in 2017-18. Lacking regular salaries or incomes, the agriculture, migrant and other informal workers would be hardest hit during the lockdown period.

Agriculture Supply Chains

COVID-19 disrupted some activities in agriculture and supply chains which are:

1. The non-availability of migrant labour is interrupting some harvesting activities, particularly in northwest India where wheat and pulses are being harvested

2. Disruptions in supply chains because of transportation problems and other issues

3. Prices have declined for wheat, vegetables, and other crops, yet consumers are often paying more

4. The closure of hotels, restaurants, sweet shops, and tea shops during the lockdown is already depressing milk sales.

5. Poultry farmers have been badly hit due to misinformation, particularly on social media, that chicken are the carrier of COVID-19.

Measures for Smooth Working of
Agriculture Sector and Supply Chains

The Government has correctly issued lockdown guidelines that exempt farm operations and supply chains. But at the same time implementation problems that lead to labour shortages and falling prices should be rectified. Keeping supply chains functioning well is crucial to food security. It should be noted that 2 to 3 million deaths in the Bengal Famine of 1943 were due to food supply disruptions but not a lack of food availability. Farm populations must be protected from the coronavirus to the extent possible by testing and practicing social distance. Farmers must have continued access to markets. This can be a mix of private markets and government procurement. Small poultry and dairy farmers need more targeted help, as their pandemic related input supply and market access problems are urgent. Farmers and agricultural workers should be included in the government's assistance package and any social protection programs to address the crisis. As lockdown measures have increased, demand has increased for home delivery of groceries and e-commerce. This trend should be encouraged and promoted during the lockdown period. The government has promoted trade by avoiding export bans and import restrictions.

Social Safety Nets as a Bridge between

Health and Economic Shock

The lockdown has choked off almost all economic activity. In urban areas, leading to the widespread loss of jobs and incomes for informal workers and the poor. Estimates by the Centre for Monitoring Indian Economy show that unemployment shot up from 8.4% in mid-March to 23% in the first week of April. In urban areas, unemployment soared to 30.9% as of April 5, 2020. The shutdown will cause untold misery for informal workers and the poor, who lead precarious lives facing hunger and malnutrition. The best way to address this urgent need is to use social safety nets extensively to stabilize their lives with food and cash. The Indian government has quickly responded to the crisis and announced a $22 billion relief package, which includes food and cash transfers. Several state governments have announced their own support packages.

The central government's relief package, called Pradhan Mantri Garib Kalyan Yojana (Prime Minister's plan for the well-being of the poor), is aimed to provide safety nets for those hit the hardest by the COVID-19 lockdown. However, it is inadequate compared to the enormous scale of the problem. Nobel Prize economists Esther Duflo and Abhijit Banerji say that the government should have been much bolder with the package's social transfer schemes. The $22 billion in spending is only 0.85% of India's GDP. This is much lower than the packages passed by the United States, European and some Asian countries. India should think bigger and be spending at least 4% to 5% of GDP. The central and state governments must spend more, even if there is a one-time hike in the fiscal deficit. Some of the additional measures needed in addition to the government package are:

Food and Nutrition Security

Government warehouses are overflowing with 71 million tonnes of rice and wheat. In order to avoid exclusion errors, it is better to offer universal coverage of distribution in the next few months. Nutrition programs like Integrated Child Development Services (ICDS), Mid-Day Meals (MDM), and Anganwadis (rural child care centers) should continue to work as essential services and provide rations and meals to recipients at home. Eggs can be added to improve nutrition for children and women. Several state governments have started innovative programs to help informal workers and the poor. For example, the Kerala government is providing meals with diversified diets at the doorsteps of households.

Cash Transfers

Unemployed informal workers need cash income support. The government has provided ₹ 500 ($6.60) per month to the bank accounts of 200 million women via the Jan Dhan financial inclusion program. But this too is insufficient. We need to have a minimum of ₹ 3000 ($40) per month in cash transfers for the next three months.

Migrant Workers

There are about 40-50 million seasonal migrant workers in India. In recent days, global media have broadcast images of hundreds of thousands of migrant workers from several states trudging for miles and miles on highways; some walked more than 1000 kilometers to return to their home villages. They should be given both cash transfers and nutritious food.

Markets and Farm Prices

With growing levels of concern, recommendations for social distance, reduced travel, avoidance of crowds, closures, and other protective practices to slow the spread of COVID-19, consumers will make tough choices about food, eating away from home, and overall spending. Dairy is prominently featured in out-of-home eating, and there may be some disruptions in food service sales impacting markets and prices. The impact of the coronavirus on the broader impact on dairy prices.

Supply Chains Slowdown and Shortages

As logistics are disrupted and efforts proceed to slow the spread of the virus. Multiple concerned industry sectors are already impacted by some products, "panic buying" created additional concern. The slowdown could impact fertilizers, fuel and other input movements.

COVID-19 – Impact on Poultry Sector

The agriculture and allied activities sector is adversely hit. In fact, the poultry sector is already being affected severely. The poultry sector which is the fastest-growing sub-sector of India's agriculture eco-system and where the country has created a foothold at the global level (India is the third-largest producer of eggs and fifth largest producer of broilers) is already facing losses to the tune of ₹ 150-200 crores each day.

COVID-19 is an unprecedented challenge for India; its large population and the economy's dependence on informal labour make lockdowns and other social distancing measures hugely disruptive. The central and state governments have recognized the challenge and responded aggressively—but this response should be just the beginning. India must be prepared to scale it up as events unfold, easing the economic impacts through even greater public program support and policies that keep markets functioning.

Food Retail

Several state governments have already allowed free movement of fruits and vegetables. Highly perishable items like vegetables and dairy have a smooth flow of goods. No or low demand for poultry products due to fake social media propaganda. Rabi season harvest for wheat, rice and pulses is already impacted due to restriction on inter-state movement of labour force and /or unavailability of Agri machinery. On account of false propaganda, the poultry sector is heavily impacted. Consumer supplies of cereals, pulses, fruits, and vegetables, dairy are largely secured as they are part of the essential list of the government. Fresh meat, seafood had an erratic supply chain as there was no clarity from the state governments about the functioning of retail shops. Brick and mortar grocery retail chains and shops were operating normally, but the shortage of staff has impacted operations. It was expected that a prolonged lockdown would result in increased demand for food supplies. Online food and grocery platforms were heavily impacted on account of unclear policy restrictions in the stoppage of vehicles. Last-mile delivery platforms were impacted too. States were working out policies for food movement which may ease the situation. Inter-state food movements were erratic and impacted prices in the next few weeks. State policies on transport might influence future traders. Edible oil prices were expected to go down in the short term due to the global downturn in demand.

Food Processing

All food-based industries were allowed to function normally. There were, however, a few issues relating to interstate movement of skilled and semi-skilled labour that needed to be sorted out. Raw material supplies were not impacted so far, and measures taken by the central government should, continue vigorously. Factories adjust to just working with less labour force and overtime also in order to meet the demand. The domestic market-based players should not have a problem. Import dependent agro chemicals, and fertiliser sector had some volatility in the long term. The industrial supply of agri-commodities is expected to be normal in the short term.

Food Exports

Major destinations such as the USA, Europe, China may continue to grapple with the virus for the next 6–8 months. Indian export-based companies, therefore, would be impacted on account of low consumer demand and port hurdles. Farm-gate prices for export-oriented commodities such as seafood, mango and grapes were crashing, and this might impact future crop prospects.

COVID-19 - Lockdown Effect on Food Imports

India ranks among the leading producers of agricultural products in the world. From a net importer of food grains in 1950s, India has emerged as a significant exporter of agricultural commodities. As per the WTO's Trade Statistics, India's share in agricultural exports and imports in the world in 2016 was 2.1 percent and 1.8 percent, respectively. India has developed export competitiveness in certain specialized agricultural products like basmati rice and castor.

Self-reliant Agriculture is Critical for the
Goal of an Atmanirbhar Bharat

Agricultural export is extremely important as besides earning precious foreign exchange for the country exports help farmers/producers/exporters to take advantage of the wider international market and increase their income. Exports have also resulted in increased production in the agriculture sector by increasing area coverage and productivity.

Agricultural trade as a percentage of agriculture Gross Value Added (GVA) has been more than 15 percent and witnessed an increase during the year 2016-17. The value of agricultural exports has increased from ₹ 215,396 crore in 2015-16 to ₹ 2,27,880 Crore in 2016-17, registering a growth of about 6 percent. The increase in agricultural exports during 2016-17 was mainly due to the exports of marine products, spices, groundnut, oil meals and fruits and vegetables. However, the share of agricultural exports in total exports of the country has marginally come down from 12.55 percent in 2015-16 to 12.30 percent in 2016-17. India's top 10 agricultural export commodities in terms of quantity and value for the last few years are:

1. Marine Products
2. Buffalo Meat
3. Rice Basmati
4. Spices
5. Rice (other than Basmathi)
6. Cotton Raw includes Waste.
7. Sugar

8. Fresh vegetables
9. Coffee
10. Groundnut.

COVID-19 lockdown very heavily impacted on food imports, production, food distribution etc., which are discussed below in detail:

Lockdown is progressed through trials and suffers. People complained about the adverse impact of its sudden announcement on the poor, farmers, daily labour, workers, etc., who are not tuned to instant information flows. The post-lockdown concerns looked much more alarming. While the objective of the lockdown was to slowdown the spread of COVID-19, the spread seemed to be gathering speed at one part of time. As the lockdown continues, India may soon stare at a bigger problem - hunger and civil disobedience. Governments at the centre and the states acknowledged that access to food was a problem and it needed attention. Despite massive relief efforts, a vast number of people were not able to access food at satisfactory levels. India had sufficient food stocks. But the unstable, corrupt and inefficient public distribution system, additionally burdened by lockdown conditions, was unable and unlikely to reach many locations and the needy. There are many concerns that have originated during the lockdown period. Some of the concerns that were recognized at that time as needing the urgent attention of policymakers, political leaders and senior civil servants, apart from scientific and technical institutions, community-based organizations, non-government organizations and the corporate sector are urgent imperatives.

Food Distribution Issues

Public Distribution System (PDS) is an important constituent of the strategy for policy, to ensure availability of food grains to the public at affordable prices, for enhancing the food security for the poor, to aid in poverty eradication and is intended to serve as a safety net for the poor whose number is more than 330 million and are nutritionally at risk. PDS evolved as a major instrument of the Government's economic PDS with a network of over 5 lakhs Fair Price Shops (FPS) and is the largest distribution network of its type in the world. PDS is operated under the joint responsibility of the Central and the State Governments. The Central and State Governments have the responsibility for procurement, storage, transportation and bulk allocation of food grains to their respective Godowns. The responsibility for distributing the same to the consumers through the network of Fair Price Shops (FPSs) rests with the State Governments. The operational responsibilities including allocation within the State, issue of ration cards, supervision and monitoring the functioning of FPSs rest with the State Governments which is the only touch point for the end beneficiary in the total Public Distribution System (PDS).

Key Factors of Production

India's food production in 1950 was limited to 50.82 million tonnes and India was known as "ship-to-mouth" country. The US Senate once refused to continue PL-480 assistance to India. From that critical position of extreme food insecurity India today produces about 295.00million tonnes (estimated) of food grains about 49% of which is produced in Kharif season, mostly dependant on rain, especially for rain fed crops.

Apart from food grains production India has shown enviable progress in cotton production, i.e., from a mere 30.4 lakh tonnes of cotton in 1950 today, it produces about 330 lakh bales of 170 kgs each as against 354.50 lakh bales estimated by it earlier. It also exports many agricultural commodities like Rice (Basmati), Wheat, Vegetables, Fruit, Tea, etc. MHA, GoI has exempted farming operations by farmers and farm workers from procurement operations etc.

The coming agricultural season, i.e., Kharif 2020 needs to be planned and envisioned in such a way so as not to disturb anti-COVID-19 operations, nor dissipate energies of communities. Every State is required to prepare a Kharif plan, which considers all factors and yet prioritises food production. It is not difficult in India in comparison with other countries. Every state needed to secure all the factors of production for the oncoming agricultural season, the rainfall being an uncertain factor. To begin with, governments should prioritise food crops by actively dissuading farmers from going to their conventional choices of crops. It should also develop confidence among farmers that this plan will get them food and a chance to earn a better income.

A local procurement plan is a necessity. This procurement should have inherently included primary processing, otherwise, the whole plan here becomes redundant. Additionally, local distribution should have the top priority, even while actions to move the surplus to other needy areas are integrated into the plan.

1. The impact is likely to be low both on primary agriculture production and the usage of agricultural– inputs such as seeds, pesticides and fertilisers.

2. Movement of migratory labour should have been allowed and facilitated for harvesting crops such as paddy, wheat, pulses during the ongoing Rabi season.

3. Insulating the rural food production areas very important in order to contain the macro impact of the virus on the food sector.

4. Crop procurement and Mandi operations are yet to be streamlined – may result in low sowing in the upcoming Kharif season and also impact the sale of agricultural inputs.

5. Long term impact will be known once the first Kharif sown reports are out

6. Procurement and marketing of Rabi season commodities will decide future prices for consumers and industries.

Urban Labour

The labour market In India is also facing a serious problem of unemployment/underemployment. A huge number of the workforce of our country remains partially or wholly unemployed throughout the year or some part of the season. This has led to problems like disguised unemployment, seasonal unemployment, general unemployment and educated unemployment. In the absence of adequate growth of employment avenues, the unemployment problem in the country is gradually becoming much more alarming day by day. Moreover, due to the policy of downsizing followed both in the public and private sector and also in government administration and services sector, the problem of unemployment is becoming much more acute. This has also been putting much pressure on the labour market of the country.

With 'urban' labour moving to their own villages, they were both strengths and weakness. Previously, most have moved out for better incomes and lack of work in villages, or because of being landless. The proposed Kharif plan should be able to include them appropriately so that their daily food needs are taken care of within the village food production cycle without being completely dependent on 'external' PDS supplies.

Further, landless families, who went out of their village for work, and small and marginal farmers, who thought urban wages could supplement their incomes, would return to their native places. It was expected that 'urban' labour would move back to villages once the lockdown is relaxed at the most opportune time for two reasons: they were sure that work in urban areas was not going to pick up in the near future and they think their hunger can be taken care of better in their own villages.

Post-lockdown, governments should have thought of village-level stocking of food because centralised, Food Corporation of India (FCI) level godowns would disrupt anti-COVID-19 operations. Local food for the local populace should have been prioritised. Economics of such prioritisation could not be a concern when hunger becomes a serious social problem. Given this scenario, both Rabi crop harvests and coming Kharif season needed more attention from the Central and State governments. In most developed countries, where the population dependent on primary sectors is low, the collapse of economies will be hard.

Short, Medium and Long-term Recommendations

1. **National Policy on Food Supply Chain during COVID-19 Pandemics:** All available evidence indicates that Food insecurity is a much more serious concern in India. Nonetheless, India has witnessed some progress in that the

incidence of severe under nutrition among children, and the incidence of certain nutritional-deficiency diseases has been declining. Chronic caloric and micronutrient deficiencies, however, remain widespread among children and adults. The most important outcomes of India's right to food movement—the enactment of the National Food Security Act (NFSA) in 2013, which has attracted considerable attention both at home and abroad regarding its potential to radically improve the food security of over 800 million Indians, therefore the policy on Food Supply Chain during the pandemic should include:

States and Centre should have classified essential food items with zero hurdle supply chain mechanism for food retail and food industries to help consumers, food industry and farmers. There should have been strict regulations against fake news propaganda impacting farmers and food processors. E.g., Poultry. Food packing industry should have been allowed as an essential category. Existing infrastructure of GST, FASTAG should have been used for smooth movement of essential food items. This would have helped in long-term stability of the food sector. Dedicated food transport corridors should have been announced pan India with no stoppage at borders. Aadhaar based approvals, passes should have been issued for a smooth supply chain of food. RBI stimulus would only help the industry in the short term.

2. **Easing Financial Stress in the Agriculture Sector:** The role and importance of investment for productivity and growth cannot be understated and relationship between capital formation and agricultural growth is well established. Gross Capital Formation (GCF) in agriculture and allied sectors in the country witnessed an increasing trend during the last three years and was 15.2 percent of GVA of agriculture and allied sectors in 2017-18. The share of public investment has increased from 11.9 percent in 2013-14 to 17.3 percent in 2016-17, while the share of the household sector declined from 85.1 percent to 79.6 percent during the corresponding period. The share of private corporations has increased marginally but is only 3 percent. Therefore, there is an urgent need to attract more corporate private sector investment in agriculture by creating an enabling environment and making necessary changes in the regulatory and institutional frameworks such as APMC Act, ECA, land tenure, contract farming, etc.

   - RBI and Finance Minister announced measures would only help the industry and the employees in the short term.
   - States must have supplied agri-inputs free of cost for the upcoming Kharif season to ensure stable food production.
   - Food Processors: Domestic and export market incentives should have been introduced for the financial year 2021 to process and liquidate inventories.
   - Agri-inputs – Priority handling of goods at the ports

3. **Support for Food and Agri Inputs Delivery Personnel:** In India every year, we distribute a huge quantity of inputs, i.e., seeds, fertilizers and pesticides as indicated below:

   - Seeds   - 372.61 lakh quintals
   - Fertilizers   - 265.91 lakh tonnes
   - Pesticides   - 5.8 lakh tonnes

   Through a network of input dealers throughout the country apart from farmer to farmer exchange/supplies. Health and life insurance for the last mile delivery personnel. Employee provident fund support should have been extended to all delivery personnel irrespective of salary amount. Registering un-employed youth as temporary staff for food and input delivery in urban and rural areas. Drone-based crop spraying should have been tried in case labour shortages arise. National Agricultural Labour force register should have been maintained connecting which should be maintained which should be connected with Aadhaar and Direct Benefit Transfer (DBT) based system. This would have helped in the rapid identification and movement of labour force to the required States for agricultural/processing operations.

4. **e-Commerce based Applications:** The Indian e-commerce industry has been on an upward growth trajectory. In February 2019, the GoI released the Draft National e-Commerce Policy which encourages FDI in the marketplace model of e-commerce. Further, it states that the FDI policy for the e-commerce sector has been developed to ensure a level playing field for all participants. According to the draft, a registered entity is needed for e-commerce sites and apps to operate in India. The government also proposed the National e-Commerce Policy, set up the lawful agenda on cross-border data flow; no data will be shared with foreign governments without any prior authorisation of the Indian government. Through its Digital India campaign, the GoI is aiming to create a trillion-dollar online economy by 2025.

   In view of the above e-Commerce-based apps should have been encouraged to help the rapid deployment of delivery personnel to avoid panic buying and restrict movement in the streets. Ensure optimal farm gate prices for domestic and export commodities and easy availability of agri labour would have ensured farmers' interest in Kharif season and therefore the food production would not be impacted. National e-Commerce policy during pandemic situation and safety net policy for healthcare personnel would have been implemented. Agriculture and allied export policy during pandemics need to be commissioned. Aadhar, PAN, passport, etc., based national arrangements were needed for the delivery of goods and services during the currency of pandemics.

2.4 Natural Resource Management

The Gross Value Added (GVA) by the agriculture, forestry and fishing sectors in India is estimated at ₹ 18.53 trillion (US$ 271.00 billion) in FY18. The Indian food industry is poised for huge growth, increasing its contribution to world food trade every year on account of its immense potential for value addition, particularly within the food processing industry. The Indian food and grocery market is the world's sixth-largest, with the retail sector contributing 70 per cent of the sales. The Indian food processing industry which accounts for 32 per cent of the country's total food market, is one of the largest industries in India and is ranked fifth in terms of production, consumption, export and expected growth. It contributes around 8.80 and 8.39 per cent of Gross Value Added (GVA) in manufacturing and agriculture respectively, 13 per cent India's exports and six per cent of total industrial investment.

Degradation of natural resources has a direct negative impact on the livelihoods of particularly the resource poor. However, experience from India shows that improvements in resource productivity per se cannot be equated with poverty reduction. As an example, the watershed development programmes have illustrated that the poor have often been excluded from accessing gains in productivity as well as related decision-making processes.

Forestry

After years of failing to protect forests through the exclusion of people from forest lands, the government has acknowledged that the inclusion of people in forest protection and management is the only way forward. Joint Forest Management (JFM), which entitles villagers to certain rights to forest produce in lieu of protection, is the main policy and strategy through which local people's participation is realised. Yet, the success of JFM is constrained by the lack of local capacity to manage forests; inadequate capacity of the Forest Department to extend appropriate support to local people; ambiguous rules and regulations; weak legal status of JFM committees, and; symbolic rather than actual participation of marginalised groups. Apart from creating a more enabling legal environment, capacity-building for both the Forest Department and community institutions are needed to improve forest management and make JFM more community-driven and inclusive.

Despite these generic problems of JFM, there are positive developments future interventions can be built upon. In Orissa the presence of a large number of self-initiated forest protection efforts; an extensive number of NGOs active in forestry; the presence of a state-level JFM committee, and; the evolution of confederations of Community Forest Management organisations, contribute to an enabling context for community forestry to progress. In addition, local control over non-Timber Forest Products following the provisions introduce in the73ʳᵈAmendment to the Constitution of India, offers an opportunity for communities to exercise greater control over minor forest produce.

Water

The rising demand for, and the poor management of water resources have resulted in increasing water scarcity in India to alarming levels. For decades, the approach to water resources has been focusing on exploitation and development rather than management including conservation. Subsidies for irrigation and electricity services have encouraged inefficient water management and overexploitation of groundwater. The extant policy, legal and institutional framework is inadequate to deal with the current problems. The plethora of government departments involved in water inhibits effective formulation, implementation and coordination of sound strategies. Lack of political will and resistance from large farmers constrain enforcement of water use regulation and tariff rationalisation despite these being recognised as important measures in the National Water Policy. Integrated Water Resource Management (IWRM), as a holistic approach to water management, is yet to gain ground in India.

However, some headway has been made towards water resource conservation and management in donor supported programmes by highlighting the need for institutional reforms and people's participation. It is now widely accepted that community-based water management should be a key strategy for analysing, planning and managing water resources in an equitable and sustainable way. The government as well as donors are increasingly looking to Gram Panchayats to perform this role. However, substantial inputs will be required to build the knowledge and capacity of Panchayats in this regard.

Despite substantial government investments in rural water supply, quality of and access to water sources remains an issue in many parts of the country. Apart from increased water scarcity, centralisation of water infrastructure and lack of local involvement and management have intensified the problem. Acknowledging the unsustainability of water supplies, the government has embarked on a sector reform programme that makes Panchayati Raj Institutions responsible for operation and maintenance. However, reviews suggest that more clarity is needed of institutional roles and strengthening of Panchayats effectively to manage the reform process. Coverage by rural sanitation is also limited in India. The key problem is perceived as low demand due to lack of awareness and failure of large-scale sanitation initiatives to develop locally suitable options.

Land Resource

Land in India is comparatively less productive than land in many parts of the world and is getting further degraded on account of the absence of effective management. The National Commission on Agriculture (1976) emphasized the importance of scientific land management planning for achieving food security, self-reliance and enhancing livelihood security. The National Policy for Farmers (2007) recommended the revival of the existing State Land Use Boards and their linkage to district-level land-use committees, so that they can provide quality and proactive advice to farmers on land use.

Between the years 1950-51 and 2007-08, land utilisation in India underwent significant changes. Lands under the net sown area, forests and non-agricultural uses, increased while simultaneously, the extent under forests, and agriculture, almost halved from 40.7% to 22.63%. Further, the per capita amount of agricultural land reduced by 67% from 1951 (0.48 Ha) to 2007/08 (0.16 Ha). Land degradation is caused by natural processes such as wind or rain and is often aggravated by deforestation and urbanisation. ICAR estimated that, as of 2010, 120 M.ha out of a total geographical area of 328.73 M.ha was affected by land degradation.

Wastelands can result from inherent/imposed disabilities such as location, environment, chemical and physical properties of the soil and financial management constraints and can be brought under vegetation cover with reasonable effort.

Degradation causes a decline in soil fertility, and alkalinity/salinity/acidity of land and water logging. Degraded lands are often used by marginal farmers and tribals, resulting in reduced productivity.

The following measures need to be taken, ensuring sustainable growth and development of the agricultural land and for the attainment of food security:

- Ensuring that all men and women, in particular the poor and the vulnerable, have equal rights to access ownership and control over land and other forms of property, inheritance, natural resources, appropriate new technology and financial services, including microfinance.

- Substantially enhancing the agricultural productivity and incomes of small-scale food producers, in particular women, indigenous peoples, family farmers, pastoralists and fishers, including through secure and equal access to land, other productive resources and inputs, knowledge, financial services, markets and opportunities for value addition and non-farm employment.

- Ensuring sustainable food production systems and implement resilient agricultural practices that increase productivity and production, that help maintain ecosystems, that strengthen capacity for adaptation to climate change, extreme weather, drought, flooding and other disasters and that progressively improve land and soil quality.

- Enhancing inclusive and sustainable urbanization and capacity for participatory, integrated and sustainable human settlement planning and management in all countries.

- Supporting positive economic, social and environmental links between urban, peri-urban and rural areas by strengthening national and regional development planning.

- Integrating climate change measures into national policies, strategies and planning.

- Promoting mechanisms for raising capacity for effective climate change related planning and management in the Least Developed Countries and Small Island Developing States, including focusing on women, youth and local and marginalized communities.

- Combating desertification, restore degraded land and soil, including land affected by desertification, drought and floods, and strive to achieve a land degradation-neutral world.

- Integrating ecosystem and biodiversity values into national and local planning, development processes, poverty reduction strategies and accounts.

- Ensuring responsive, inclusive, participatory and representative decision making at all levels.

The link between Poverty and NRM

Despite rapid urbanisation and increased livelihood diversification, more than 60 per cent of India's population still depend on agriculture for livelihoods. The nexus between poverty and environmental conservation remains strong. Land degradation is a key issue affecting resource productivity. It is estimated that about one-third of the soil in India has been affected by erosion. This has a direct impact on agricultural productivity and hence food production, especially for the resource poor farmers living with marginal landholdings. The area declining under forest cover has now been arrested, but the volume and density of forests have been reduced, causing scarcity of valuable forest produce, important for the livelihoods of poor in many regions (TERI, 1998).

Overexploitation of surface and groundwater threatens the quantity, reliability and quality of water availability. Water is rapidly becoming a scarce resource. Growing scarcity and competition for water poses a major threat to advances in poverty reduction by limiting productive as well as consumptive aspects of livelihoods and well-being. If this trend is not reversed, it is believed that an increasing number of poor people may find more difficulty in securing access to water than securing access to food, primary health care, or education

(Barker et al., undated). Access to safe water supply and sanitation means a lot for economic and social well-being in India. Fetching water, a task that is often placed on women and children, can be labour intensive and exhausting. Finding a place for defecation can be especially time-consuming for women. Illness, or even death, caused by water-related diseases can place a heavy strain on a household's livelihoods by loss of income, cost of medicines and time spent on care. Despite significant improvements in drinking water supply and sanitation during the last decades, environmental factors are still responsible for many diseases caused by inadequate hygiene, water contamination, and unsafe supplies. More than 163 million Indians still do not have access to safe drinking water and nearly 1.5 million children die each year due to water-borne diseases higher than the population of Russia – do not have access to safe drinking water. A report by the United Nations says that more than three million people in the world die of water-related diseases due to contaminated water each year, including 1.2 million children. In India, over one lakh people die of water-borne diseases annually.

There have been many initiatives to combat degradation and increase resource productivity, which the poor have been able to benefit from in various ways. Large-scale irrigation has played a major role in economic growth in India by increasing food production and providing employment. In many parts of India, low-cost tube well irrigation has stimulated agricultural production for small scale farmers and brought many households out of poverty. Outside high potential irrigation areas, soil and water conservation alone has increased agricultural productivity and enabled new and additional crops to be grown. Coupled with minor irrigation, productivity levels have in some semiarid areas dramatically improved household food production and security. Increased work on the land has generated employment opportunities for poor people. Forest protection activities involving communities have increased the availability of forest produce.

However, in addition to the question of sustainability of many of these initiatives, it is increasingly being recognised that natural resource management per se may not create sustainable livelihood outcomes for the rural poor. In fact, scattered evidence at times suggests experience to the contrary. As an example, watershed development programmes in India have largely rested on the assumption that improvements in the resource base would automatically lead to reduction in poverty. However, not only has it become evident that these programmes have been biased towards the landed (in terms of increase in water availability and land productivity), but also that they can work against the interests of the landless by reducing their access to common lands in the name of protection. Livestock belonging to the poor have been banned from entering newly planted public and private plantations. Increased productivity has resulted in encroachments onto public lands for agriculture and the poor being denied access to grazing. In some cases, lack of access to grazing land has forced the landless to sell their livestock. Where aquifers have been recharged and water availability increased, land-rich households have been able to capitalise on the water to increase their production, at times over-exploiting water resources, especially groundwater, to negate gains. Local committees established in watershed programmes often reflect social and power relations in villages, limiting the influence of the poor.

Clearly the link between improved resource productivity and poverty reduction is far from automatic. Perhaps one of the biggest challenges of future NRM programmes will be to contribute to poverty reduction by creating sustainable improvements in the livelihoods of the poor. This will require better targeting and supporting positive directions of change with regard to resource access and equity. Poor people must be included to a greater extent and have a larger say and share in NRM programmes. It has been demonstrated that support to a geographical area with a high incidence of poverty or a strategy based on 'poverty inclusion', are inadequate approaches to combat poverty.

Future strategies must be based on a situation-specific analysis of resource use at the household and community level, and an understanding of how this is related to local as well as external social and economic institutions and processes. Thus, for NRM strategies to create pro-poor outcomes, many will need improved understanding including:

- Identification of the poor and stabilising the case for their inclusion. Determining how their livelihood portfolios and strategies linked to different types of natural resources? How can increased availability of, for example, water or forest produce improve their asset base and contribute to enhance their livelihoods and well-being.

- Recognising the barriers that reduce poor people's claims and access to resources and examining how they can be dismantled

- Identifying opportunities exist for enhancing poor people's inclusion and power in community and political processes related to resource management.

- Establishing the link between the local policy and wider policy issues and determining how poor people's rights to water be negotiated and improved within that. The scope for change in policy and practice also needs to be investigated.

To understand whether an NRM intervention will have desirable livelihood outcomes for the poor requires a framework for poverty analysis. It is generally agreed that poverty is multi-dimensional and dynamic and originates in a set of interlocking factors that reinforce ill-being in individuals, households and communities. The Department for International Development (DFID) of the government of UK draws extensively on the Sustainable Livelihoods (SL) approach as a framework for analysing its rural development programmes in India. Although there are reservations about the unqualified application of the SL approach as a blueprint for poverty analysis, it does nevertheless offer a starting point for unpacking micro and macro links of resources and livelihoods by looking at households, and also by drawing attention to institutional and policy structures. Increased attention is also being accorded by various tangible and non-tangible dimensions of vulnerability including economic vulnerability; social vulnerability; vulnerability arising from lack of knowledge and information; and vulnerability arising from natural disasters. In many instances, the vulnerability concept can provide some answers to not only who *is* poor today, but also who is likely to *become poor / poorer*. Studies are beginning to emerge in India along these lines describing people's experiences of poverty and categorising the problems of different groups of poor and how their needs can be better met at the programme and policy level.

Ground Water for Sustainable Food Security

Water is the key input for agriculture. The choice of crops by farmers as well as nature depends largely on the extent of water availability. Seawater constitutes 97 per cent of the world's water resource. Nearly 25 per cent of the global population lives near the coast. Therefore, it is essential that we develop technologies for bio-saline agriculture. Among the other sources of water, ground-water is becoming increasingly important. Surface water resources such as tanks, lakes, reservoirs, and rivers are equally important from the point of view of clean technology development and adoption. Groundwater in eastern India contains arsenic and this can be removed by bioremediation techniques. Drip and sprinkler irrigation is another notable clean technology. Protected agriculture, particularly horticulture, provides an opportunity for linking drip irrigation with organic farming. Organic agriculture helps to improve soil health from the point of view of the physics, biology and chemistry of the soils.

Ground Water Sanctuaries

India will be required to produce more and more from less and less land and water resources. Alarming rates of ground water depletion and serious environmental and social problems of some of the major irrigation projects on the one hand, and the multiple benefits of irrigation water in enhancing production and productivity, food security, poverty alleviation, as mentioned earlier, are too well known to be further elaborated here: Water availability per capita has declined from 5000 cubic metres (m^3) per annum in 1950 to around 2000 m^3 in 2019 and is projected to decline to 1500 m^3 by 2025 leading to far less water availability for agriculture. Agriculture is the biggest user of water, accounting for about 80 percent of the withdrawals. There are pressures for diverting water from agriculture to other sectors. On the other hand, inefficient and dilapidated canal irrigation systems have led to a spurt in groundwater development. India is the largest user of groundwater in the world with over 60 per cent of irrigated agriculture and 85 per cent of drinking water supplies dependent on aquifers. A study has warned that the re-allocation of water out of agriculture can have a dramatic impact on global food markets.

The input subsidies and commodity price policies favour crops like paddy that are water intensive. While it is good to use water efficiently in agriculture, a case is being made out of where water from this sector to be reallocated to other economic sectors like industry. The water use may be reduced by as much as 23 per cent by 2022 and this may result in drop of yields of rice, in particular, its price rise and reduced food grain availability for poorer sections. Several States offer free electricity to draw out groundwater for irrigation. Some, in fact, offer a 24x7x365 uninterrupted supply of power in addition. However, erratic power supply forces farmers to rely on diesel-run generators, which add to the cost. With the water table going down drastically on account of overexploitation, expenditure on re-digging of bore wells and higher capacity motors is also unavoidable. To deal with this crisis, aquifer recharge and rainwater conservation through community ponds and recharge wells needs to be promoted with the involvement of local bodies.

Ground water sanctuaries need to be created in every drought-prone district. Large aquifer areas on government land should be identified and ground water from these areas should be preserved. Water from such 'sanctuaries' (to use the expression of Dr. M.S. Swaminathan), should be drawn only during periods of drought, and then only for drinking purposes for human beings and cattle. Arrival of monsoons, the government needs to ensure that these sanctuaries get recharged to preserve water for the future. Such sanctuaries should be made in every block panchayat so that during severe water scarcity period, people can use this.

Every drought needs to be viewed as an opportunity to prepare better for the future. Once in every three to four years, rainfall is going to be erratic. For that the government should prepare and train farmers accordingly. Dry farming needs to be encouraged. Whenever rainfall is good, the government needs to supply quality seeds and fertilisers to encourage farmers to produce more food grains. This will compensate for the shortage of food production during a drought year.

Policy reforms are needed in the future including establishment of secure water rights to users, decentralization and privatization of water management functions to appropriate levels, pricing reforms, markets in tradable property rights, and the introduction of appropriate water-saving technologies.

Water Woes of Farmers in Bundelkhand

For the last 25 years, Haritika a NGO started by Avani Mohan Singh has solved water issues of farmers in Bundelkhand, increased water availability, and augmented cultivable land while helping with infrastructure development to make villagers living in the backward districts self-reliant. Bundelkhand might become water scarce by 2030 is a serious problem that the agricultural community in this area is battling. Time, resourcefulness, and intervention can avert disaster, and the ideological Avani Mohan Singh has been tireless in addressing this grave issue. The local hero wanted to address Uttar Pradesh's perennially parched region's cry for help, and show others the way.

Avani has been working to solve issues affecting the rural poor. From interventions on natural resources management, and provisions of infrastructure to make villagers living in the backward districts of Bundelkhand region self-reliant, Avani and Haritika's efforts are already making a difference.

A land without water

A recent study, Vision Document for Bundelkhand, commissioned by the UP government, cautioned that Bundelkhand might become water-scarce by 2030. The region between River Yamuna and the northern part of the Vindhyan plains, comprises seven districts - Banda, Chitrakoot, Hamirpur, Jalaun, Jhansi, Lalitpur, Mahoba, etc. A population of 78 lakh relies on perennial rivers like Yamuna, Ken, Betwa, Sindh, and Pahuj for its water needs. Over 75 percent of the population depends on agriculture as a primary livelihood, with 96 percent of the income generated from agriculture and livestock collectively. The study was conducted by the Church's Auxiliary for Social Action (CASA) and Jan Kendrit Vikas Manch (a network of NGOs in Bundelkhand).

"Being a rain-fed area, it is significantly dependent on rainfall, which has progressively decreased over the past decade leading to severe droughts". According to reports, there has been a 60 per cent decline in rainfall in the past five years, and this concerned Avani and led him to start Haritika and address Bundelkhand's water issues. "The organisation works as a project implementation agency, and has, since 1994, done hundreds of water-related projects and touched the lives of seven lakh people in 500 villages in Madhya Pradesh and Uttar Pradesh".

Quenching Farmers Thirst

Avani is very proud of Haritika's key role in the Integrated Watershed Project, which the Coca-Cola India Foundation launched in 2013. It involved rainwater harvesting and water conservation in the Bundelkhand region. The project is a joint effort of Coca-Cola India, International Crops Research Institute for the Semi-Arid Tropics (ICRISAT), and Haritika along with national partners National Research Centre for Agro-Forestry (NRCAF), district administration, and the local community."It helped improve the economic conditions of the residents of Parasai-Sindh region by increasing water availability and cultivable land," Avani says.

"A series of eight check dams were constructed, which developed 125,000 m^3 of storage capacity. After completion of the Parasai-Sindh watershed, the groundwater table on an average increased by around two to four metres as compared to the non-intervention (or control watersheds) stage. This has increased cropping intensity by 30-50 percent, especially during the post-monsoon season," he adds.

Thanks to the project, farmers have started harvesting wheat ranging from 3,500- 4,000 kg per hectare (on an average), which has significantly improved income and livelihood. Farmers also shifted from low-water requiring crops to high-value crops.

"Earlier, open wells hardly supported even one to two hours of irrigation during the Rabi season due to the low water column. Farmers (mainly women) spent 10 to 15 days or 40-50 hours irrigating one-hectare of wheat crop. Now, a majority of the wells support round-the-clock irrigation and it is done in a day (15-20 hours)".

2.5 Role of the Corporate Sector

India has seen the growth of contract farming with companies directly getting into a contract with the farmers and providing them a "fair price" for their produce. This is leading to corporate farming as a natural extension. A few initiatives have been taken in this regard with big corporate houses stepping in. The introduction of this type of farming and the involvement of the corporate might provide the much-needed acceleration to the sector.

Corporate Farming Case

Jamnagar Farms Private Ltd. – Subsidiary of Reliance Industries

Jamnagar Farms, the 1700 acre agri-forestry, agri-horticulture farm set up in arid land near the Reliance Industries Ltd. (RIL) refinery, uses the recycled waste water from the refinery for irrigation purposes. Though initially taken up as part of the plan to improve the environment, the company has a profitable proposition and has prepared a roadmap for expanding farm production in Andhra Pradesh, Karnataka and Maharashtra. Fast forward to 2020, that orchard, now known as 'Dhirubhai Ambani Lakhibag Amrayee' is Asia's best mango orchard with more than 1.3 lakh plants of over 200 species. Within the last 6-7 years there has been a total transformation of the barren wasteland into lush green countryside. Initially there was a lot of scepticism about the success of mango at Jamnagar owing to high velocity winds, salinity in water and soil. However, with the adoption of appropriate technology mango has become a success story. One cannot find such a large scale scientifically managed successful mango plantation anywhere in this part of the world.

The mangoes produced in that orchard are of excellent quality and are widely exported at a global level. Reliance grows 127 varieties of mango in a 600-acre green belt at its Jamnagar refinery complex, looks to beat Israel and Brazil in productivity. Jamnagar Farms currently owns Asia's biggest mango orchard.

2.6 Role of Biotechnology in Agriculture

Biotechnology offers multiple innovative techniques to develop high-yielding crops that can counter the biotic and abiotic stress associated with Indian agriculture. Unfortunately, the debate on biotechnology often gets limited to GM crops, whereas the reality is different—we need to look at the complete array of solutions provided by biotechnology and use it in a more comprehensive manner.

Bio agriculture yields better results than traditional techniques while maintaining the stability and fertility of the soil. High yielding seeds significantly enhance the productivity potential and provide resistance from adverse environmental stress such as drought and salinity. They are particularly effective and relevant for a country like India that suffers from water scarcity and drought every year. High-yielding seeds also protect crops from diseases and insects.

Plan for the application of agricultural biotechnology to bring about transformative changes in the agro sector.

Some of the recommendations include:

- Introduction of Bthgenes, which are already used the world over, in India.

- Release of genetically modified crops that have cleared all the safety tests for commercial use or for large scale field trials in partnership with different states.

- Establishment of beneficial international collaborations including access and acquisition of IPR-protected technologies in the areas of genome editing, genetic modification, marker assisted selection (MAS) and breeding for post-harvest losses.

- Efforts to disseminate correct information about new technologies to the public to dispel myths.

2.7 Research, Extension, Metrological Services

Agriculture Research System in India

The public research system in India is led by the ICAR, which has 5 multidisciplinary national institutes, 45 central research institutes, 30 national research centres (NRCs), 4 bureaux, 10 project directorates, 80 all-India co-ordinated research projects (AICRPs)/networks and 16 other projects/ programmes. In addition, there are 4 deemed universities, 64 state agricultural universities (SAUs), and 3 Central Agricultural Universities which operates through 313 research stations. AICRPs are the main link between the ICAR and the SAUs. The number of centres involved in the AICRPs is about 1,300, of which about 900 are based in agricultural universities and 200 in the ICAR institutes. The ICAR has also Zonal Research Stations (ZRSs) and 200 sub-stations. The National Academy of Agricultural Research Management (NAARM) is another institution under ICAR to conduct research and training in agricultural research management. The ICAR has also established 8 Trainers' Training Centres (TTCs) and 611 Krishi Vigyan Kendras (KVKs) at the district level as innovative institutional models for assessment, refinement and transfer of modern agricultural technologies.

In addition, there are 23 general universities under the University Grants Commission (UGC), involved in agricultural research. Several scientific organizations such as the Council of Scientific and Industrial Research (CSIR), Bhabha Atomic Research Centre (BARC), National Remote Sensing Agency (NRSA), Ministries and government departments such as Ministry of Commerce, Department of Science and Technology, Department of Biotechnology, Department of Ocean Development, and more than 100 private and voluntary organizations and more than 105 scientific societies are involved in the agricultural R and D and they form the part of the national agriculture research system of India (Vision 2020, ICAR). This extensive agriculture research infrastructure not only conducts agriculture research but are also responsible for educating and providing extension services to the farmers.

Over the past twenty-five years global production of agriculture has been outpaced by unprecedented population growth. The population of the world is expected to rise from 6 billion to 8 billion by 2020 and due to this food and nutritional security is becoming an upcoming challenge for the developed countries all over the world. For most of the developing countries opportunities and growth are getting constrained because of degradation and over-exploitation of natural resources. Positive policy support, greater public funding and dedication of farmers in work have contributed a lot for increasing agricultural, animal and production of fishes in India. A golden revolution in horticultural crops is on the horizon through a significant increase in production. This golden revolution is resulting in the distinct gaining of actual consumption of households and qualitative improvement in human diets.

The agriculture and allied sector of India focuses on food and nutritional securities, sustainable development and poverty alleviation. Its major role is to conserve its natural resources, to maintain its biological wealth and to accelerate agricultural growth for present and future context. It focuses on the development of novel technologies by providing problem solving knowledge. The sectors face several challenges that are arising from both supply and demand-driven perspectives. The sector of agriculture can be made more sustainable and remunerative by evolving mechanisms for enhancing the capacity of all stakeholders.

The Indian Council of Agricultural Research (ICAR) plays an important role in the development of Indian agriculture. This council's major role is to promote excellence in higher education in agriculture. It is involved in developing areas of science and technology along with its scientists that are internationally acknowledged in their respective fields. It follows some of the mandates such as to promote and co-coordinate education in agriculture, to provide and promote consultancy services in the field of education and to solve problems occurring in areas related to agriculture. ICAR focuses, among others the key areas including:

Promoting innovations and in improving human resource capacity, strengthening institutional capacity for attaining sustainable development, achieving global competitiveness, enhancing the national wealth of natural resources and diversity by acting as a catalyst. In promoting measures for adapting and mitigating climatic challenges, finding out mechanisms for effective drought and flood management and the impact of fostering repositories for making a sustainable utilization of resources.

Agricultural Extension - Institutional Mechanism

The extension services will be implemented through the institutional mechanism as given below:

State Level

The State level Sanction Committee (SLSC) set up under Rashtriya Krishi Vikas Yojana (RKVY) is the apex body to approve the State Extension Work Plan (SEWP) which will form a part of the State Agriculture Plan (SAP). The State Level Sanction Committee (SLSC) will be supported by the Inter Departmental Working Group (IDWG). IDWG is responsible for day-to-day coordination and management of the Scheme activities within the State.

State Agricultural Management and Extension Training Institute (SAMETI) is the State level institution catering to the training and human resource development needs of extension functionaries and support SAMETI to create essential infrastructure. The State Nodal Cell (SNC) consisting of the State Nodal Officer and the State Coordinator (along with supporting staff) will ensure timely receipt of District Agriculture Action Plans (DAAPs), formulation of State Extension Work Plan (SEWP) duly incorporating Farmers' feedback obtained through State Farmer Advisory Committee (SFAC) and its approval by the SLSC. The SNC will then convey the approval and monitor the implementation of these work plans by SAMETIs and Agriculture Technology Management Agencies (ATMAs). The SAMETIs will draw-up and execute an Annual Training Calendar for capacity building of the Extension functionaries in the State. While doing so, the SAMETI will check duplication and overlapping of training content, training schedule as well as trainees. Farmers Advisory Committees at State, District, and Block level to advise and provide inputs to administrative bodies at respective levels.

District Level

Agriculture Technology Management Agency (ATMA) is an autonomous institution set up at the district level to ensure the delivery of extension services to farmers. ATMA is responsible for coordination and management of agricultural extension related work in the District. ATMA Governing Board (GB) is the apex body of ATMA which provides overall policy direction.

ATMA GB will be assisted by the District ATMA Cell comprising PD ATMA, Dy. PDs and Staff in the discharge of its functions. ATMA Management Committee (MC) is the executive body looking after the implementation of the scheme. District Farmers Advisory Committee (DFAC) is a body to provide farmer's feedback for district-level planning and implementation. With dedicated staff provided for the ATMA, it will continue to be the district level nodal agency responsible for the overall management of the agriculture extension system within the district, including preparation of Strategic Research and Extension Plan (SREP).

ATMA is a multi-agency platform with an emphasis on procedural as well as institutional reforms, leading to effective extension delivery.

Strengthening and reforming ATMA to meet Doubling Farmers' Income (DFI), challenges are:

1. Lack of convergence among flagship programmes/schemes under agriculture and allied sectors.
2. Poor participation and commitment on the part of senior officials.
3. Lack of clear understanding about the principles and practices of ATMA.
4. Untimely release of funds from State to ATMA (district unit) and block units.
5. Diversion of ATMA functionaries into non-core activities.
6. Capacity building of extension functionaries was not sustained.
7. Inadequate infrastructure support to extension.
8. Attrition ridden and unstable contractual manpower; also allied activities like animal husbandry and fishery sector did not receive required manpower.
9. Absence of effective monitoring mechanism at district level through the mandated ATMA Governing Body (GB).
10. Extension services need to be made more outcome oriented, with a balanced emphasis on both production and post-production activities.
11. Farm school concept of Farmer-to Farmer extension not fully operationalized.
12. Poor quality of interaction among members of Farmers Advisory Committee (FAC).
13. Lack of integration of agri-entrepreneurs with ATMA activities.
14. Non-evolution of effective Public-Private Partnership (PPP) models.
15. Quality of Strategic Research Extension Plan (SREP) less than desired.
16. Poor efforts in the promotion and sustenance of farmers groups (FIGs, CIGs, SHGs, FPOs, etc.).

17. Ineffective linkages between Agricultural Technology Application Research Institutes (ATARIs) and SAMETIs at the state level and ATMAs and KVKs at the district and below level.

Block Level

At the Block level, two bodies viz. Block Technology Team (BTT) (a team comprising officers of agriculture and all line departments within the block) and Block Farmers Advisory Committee (BFAC) (a group exclusively consisting of farmers of the block) shall continue to function jointly (with the latter providing farmers' feedback and input). BFACs shall represent Farmer Interest Groups (FIGs) / FOs existing within the block on a rotation basis to advise the BTT. The Block ATMA Cell consisting of these two bodies, Block Technology Manager and Subject Matter Specialists will provide extension support within the Block, through the preparation and execution of Block Action Plans (BAPs).

Organisational Structure of Atma Scheme

The Organisational structure at various levels has been depicted in the following diagram:

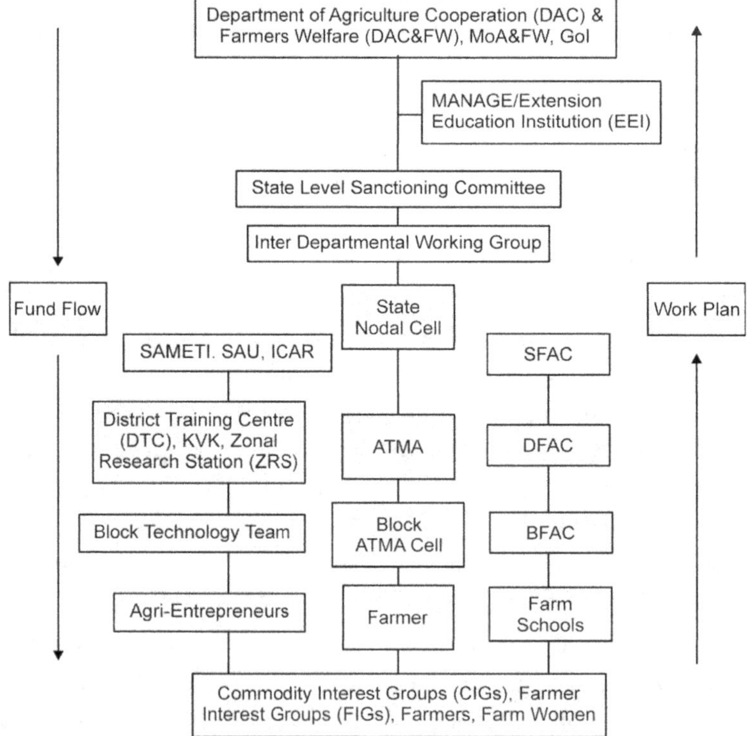

Source: Compilation by the author

Village Level

The Farmer Friend (FF) will serve as a vital link between the extension system and farmers at the village level (one for every two villages). The FF will be available in the village to advise on agriculture and allied activities. Innovative support through FF for every two villages. The FF will mobilize farmers' groups and facilitate dissemination of information to such groups, individual farmers and farm women directly through one to one interaction individually or in groups and also by accessing information/services on behalf of farmers as per need through Common Service Centres (CSC) / Kisan Call Centres (KCC). Agri-entrepreneurs will supplement the efforts of extension functionaries by making quality inputs available to the farmers and by providing them critical technical advice. Farm Schools will serve as a mechanism for the farmer-farmer extension at 3 to 5 focal points in every Block.

Decision Oriented Information System for Farmers

The Kisan Call Centres (KCC) (Farmer Call Centres) scheme has been launched as an innovative and modern scheme of the government for expeditiously delivering extension information and support to the farmers, using the vast telecommunication network which has grown remarkably. It helps overcome the handicaps of the traditional personal extension system which is often inadequate in meeting the pressing queries and demands for the latest information from the farmers. The KCC scheme was launched in 2004 by the Ministry of Agriculture and Farmers Welfare, GoI.

To make correct decisions on various critical matters, farmers frequently need information and advice on many different technical and economic aspects. The information helps them to make correct decisions on matters such as the right crop and variety to plant, the correct inputs to apply to solve problems, and the right practices to follow so as to manage their farms successfully and achieve the best productivity and returns. Inadequate and imperfect information leads to poor decisions, poor farm performance, and in the worst cases even to crop failures and suicides. Systems to provide good andup to date information and knowledge to the farmers are therefore extremely crucial for their productivity and livelihoods as well as the performance of the agriculture sector.

The modern management approach to designing a good information system focuses on the main decision-making needs of the farm managers. The approach first identifies the key decision-making needs for best achieving the objectives of the farm. Then, in order to make these key decisions well, it identifies what key information will be required.

This includes not only "what", but also "when", "where" and "who" of the information. Then, squarely based on this examination, a tailor-made information system is designed and implemented, which would most effectively and directly provide the information when and where it is needed. The result is an information system which directly leads to better decision-making and performance.

Agriculture Knowledge and Innovations

Knowledge and innovation have a key role to play in helping farmers and rural communities meet substantial challenges. These include ensuring long-term food and nutrition security, bolstering environmental care and climate action and strengthening the socio-economic fabric of rural areas. Although agricultural research delivers new knowledge and there is already a substantial amount of knowledge available to answer these challenges, it tends to stay fragmented and insufficiently applied in practice. Moreover, the agricultural sector itself has considerable and under-used innovation capacity. On average, twenty years separate the start of research from the mainstream application of its outcomes in agriculture. The insufficient or too slow uptake of new knowledge and innovative solutions in farming, in particular by small and medium-sized farms, hampers a smooth transition towards more sustainable agriculture as well as the farm sector's competitiveness and sustainable development.

Drones for Agriculture

Information and communication technologies (ICTs) are playing an increasing role in addressing problems faced by agriculture. The challenges faced by agriculture from climate change alone are enormous, and the need for the farming communities to adapt and become resilient is key to feeding the world's growing population. Harnessing the growth and transformative potential of ICTs provides a tremendous platform not only for addressing some of these challenges, but also for accelerating efforts to achieve the Sustainable Development Goals (SDGs) by 2030.

The use of unmanned aerial vehicles (UAVs), also known as drones, and connected analytics has great potential to support and address some of the most pressing problems faced by agriculture in terms of access to actionable real-time quality data. Goldman Sachs predicts that the agriculture sector will be the second-largest user of drones in the world in the next five years. Sensor networks based on the Internet of things (IoT) are increasingly being used in the agriculture sector to meet the challenge of harvesting meaningful and actionable information from the big data generated by these systems.

Potential of Drone Technology to Alleviate Agriculture Problems

The utility and benefits of drone application in agriculture are well documented. The real-time nature of information and its precision will be the key driver for agricultural development. With rapid adoption and continuous innovations, the technology will become more and more accessible to the common farmer. However, adequate training and awareness is a must for its deeper penetration into the rural masses. The future of UAVs in precision agriculture comes down to farmers being ready and willing to try out the technology for themselves. Getting involved now helps farmers acquire an understanding of the tremendous potential of drones and also allows them to determine their own way forward.

Agriculture Input Subsidies

Input subsidies have been given to promote the use of subsidised inputs and practices to get higher productivity and production, contribute positively to farmers' income and to promote sustainable use of natural resources. Though subsidies have played a significant role in raising agriculture output and farmers' income, it is believed that subsidies are not being used in an equitable and efficient manner, and in some cases they are having an adverse effect on natural resources and the long-term sustainability of agriculture production. There are also reports of excessive and indiscriminate use of some inputs and water and misuse of subsidies for non-targeted purposes. Therefore, ways and means are being discussed to pay subsidies directly into the accounts of farmers through the Direct Benefit Transfer (DBT) route rather than supplying input at subsidised prices. Suggestions also include merging all kinds of subsidies into one pack and distributing the total subsidy amount to farmers on a per acre basis. This requires a precise estimate of various subsidies being given by the Centre and states and proper understanding of the nature of subsidies.

Success Story of Y. Jagdeesh Reddy, who <u>also</u> received <u>the</u> Innovative Farmer Award, IARI, New Delhi in 2019:

Some two and a half decades ago, Reddy quit his polytechnic studies as a teenager and took up farming as a profession, with a resolve to make a difference. He started tilling his 25-acre land at his native village Nalagampalle and the nearby Moghili in Bangarupalem mandal in Chittoor district.

Like any other farmer, he had used chemical fertilizers and pesticides for almost 15 years. But, his venture gathered steam after he bid a goodbye to the use of inorganic compounds in cultivation and now, his name is synonyms with the very mention of 'natural farming' in the district. In recognition of the relentless pursuit of the goal to liberate farm produces from chemical fertilisers, the Indian Agriculture and Research Institute (IARI) in New Delhi conferred the 'Innovative Farmer' award in 2019.

"Inspired by the lectures of agriculturist and Zero Based Natural Farming (ZBNF) pioneer Subhash Palekar, I practised natural farming on a small stretch land for the first time in the 2009 kharif. I used chemical fertilizers on the rest of my land as I was experimenting," he says.

For three years, Reddy recalls, there was not much yield from natural farming while the chemical fertilizers had the upper hand. "In 2012, I had outstanding results in terms of healthy growth of paddy and sugarcane crops, which eventually ensured good yield. Once for all, I stopped using chemical fertilizers".

Trade secret

Explaining about the farming methods, Reddy says, "The dung and urine of native breed cows are what I am using as fertiliser. I use leaves of ten tree species to prepare a special concoction which I mixed with the dung and urine. I prepare Jeevamrutham, Ghanamrutham and Akula (leaves) Dhravam, which are used as fertilizer, insecticide and pesticide. This method augers well with the growth of useful worms and bacteria in the field."

Reddy has also done several experiments with growing inter-crops with paddy, mango and sugarcane. Throughout the year, his fields remain green with one crop or the other. The groundnut oil produced from Mr. Reddy's farm has takers from several northern cities. "Growing multiple crops simultaneously acts as an insurance if one crop fails," he says.

At a time when farmers across Rayalaseema are worried over the drought, Reddy appears relaxed. "Natural farming methods have proved that it can beat drought conditions as the soil moisture is preserved. Moreover, avoiding the use of chemicals leads to a tremendous reduction of heat, particularly in the summer," he explains.

Healthy produce

More than business, Reddy says his ambition is to make farmers shun chemical farming. "Natural farming will protect a farmer from huge loss for sure. It also helps save on the huge expenditure involved with chemicals. More importantly, the crop you produce is healthy, away from the risks of life-threatening ailments," he says.

Reddy's farm has now become a laboratory, attracting hundreds of students every week to have a look at the natural farming methods.

CHAPTER - 3

SUSTAINABLE GROWTH AND EQUITABLE DEVELOPMENT

While women are the backbone of the rural economy and the invisible farmers, they are often ignored indecision making, despite their close involvement with both farming and marketing activities, often directly. Yet, they receive but a fraction of the land, credit, inputs, agricultural training and information compared to men. Gender equality is a social responsibility that needs to be put into practice sinerely and with a sense of purpose. Technological empowerment, unmediated control and ownership of land, enhancing the management skills and knowledge of women in agriculture are ways by which the scourge of discrimination can be addressed..

With an increasing density of population, the size of farms is only declining apart from being subjected to the additional challenge of fragmentation. Some of the key challenges are an investment in agriculture, conservation, improved yields through better farming practices, access to market, availability of institutional credit, increasing the linkages between agricultural and non-agricultural sectors need urgent attention. An optimal solution to achieve efficiency in agriculture lies in mobilising the farmers into farmer producer organisations (FPOs), including cooperatives of various nature. Need to promote sustainable and equitable agriculture and rural prosperity through effective credit support, related services, institutional development and other innovative initiatives across the agricultural value chain.

This chapter examines the need for centre-staging roles of women, youth, community-based organisation and non-governmental organisations, scaling up successful models such as contract farming, food value chain – storage, food processing, marketing, the role of the corporate sector, role of biotechnology in agriculture. This would help the farmers realise stable returns of income from agriculture.

3.1 Centre Staging Role of Women, Youth, CBOs, and NGOs

Conventionally, the period from adolescence to middle age is termed as youth. The global population is projected to reach 9 billion by 2050, by which time the number of young people (aged 15 to 24) is expected to increase to 1.3 billion, accounting for almost 14 percent of the projected population, most of them to be born in the developing countries of Africa and Asia, where more than half the population still lives in the rural areas. By 2020 India is expected to have a 34.33% share of the youth in the total population, a huge reserve of human resources.

The involvement of the youth is very important for the sustainable growth and development of the agriculture sector. The majority of the farmers in India are more than 35 years and largely unreceptive to new technologies and practices.

A study showed that about 50% of the farming youth perceived social media as being helpful in getting new information related to farming, and that it saved time and money. Most of them knew about the Krishi Rin Mochan Yojna (KRMY). The study also showed that awareness levels about the programmes/policies related to the youth and the agriculture sector-wise very low in the farm and non-farm youth, largely on account of lack of awareness and low levels of education. There is, thus, a need to improve the education and awareness levels of the youth through appropriate modes.

The Indian Agriculture Scenario and Gender Imperatives

According to a government survey, 75 percent of rural women are farmers. Moreover, evidence suggests that women carry out most of the back-bending tasks on the field, that is, from preparing the land, selecting seeds, preparing and sowing to transplanting the seedlings, applying manure, fertilizers, and pesticides, and then harvesting, winnowing and threshing.

According to the 2013 Oxfam India report, women farmers produce 60 to 80 percent of food and 90 percent of dairy products. However, women have poor access to extension services and less than 13 percent of Indian women own land. Working as casual labourers, women's wages were only 69 percent of male wages. Yet, reports and studies have shown that women hold only 12.8% of operational holdings in India–lower than the 17% in China–over an area constituting a 10th (10.3%) of the total area of India's operational holdings, according to the Center for Land Governance index. Furthermore, women are often under-represented in rural organizations and institutions and are poorly informed regarding their rights. This prevents them from having an equal say in decision-making processes, and reduces their ability to drive collective action, such as membership of Primary Agricultural Cooperatives Societies or Water User Associations. For example, women comprised 7.5 percent of the total membership of the cooperatives in South Asia. As a result, two aspects remain critical in ensuring competitive agricultural growth trends for the future: the availability of sufficient and sustainable sources of water; and gender responsive solutions that enable more women to access finance, technology, information and market opportunities.

Figure 3.1 Women in Agriculture in India at a Glance

Source: Agricultural Statistics, GoI

The Indian agricultural sector can realize its growth potential through gender-inclusive strategies and better water resources management. The existing gender gap prevents women from effectively contributing to agricultural productivity improvements. This calls for the initiation of a policy dialogue and targeted stakeholder interventions.

Reshaping the Future in Agriculture with Youth

The global population is projected to reach 9 billion by 2050. Rural youth continue to face challenges related to unemployment, underemployment and poverty. Despite the agricultural sector's ample potential to provide income-generating opportunities for rural youth, challenges related specifically to youth participation in this sector persist more and, importantly, options for overcoming them are not extensively documented. Furthermore, statistics on rural youth are often lacking, as data are rarely disaggregated by important factors such as age, sex and geographical location.

India accounts for a substantial share of the world population. By 2010, India accounted for 17.8 per cent of the world population, an increase of 2.7 per cent in its share since 1970. This growth is projected to continue even by 2030. India's share in the decennial addition to global population increased from 18.13 per cent during 1970-1980 to 22.87 per cent during 1990-2000 and is projected to decline to 18.69 per cent by 2020-2030.

The population of India is projected close to 1.380 billion or 138 crore people in 2020 as against at 1.366 billion or 136.6 crore people in 2019 with 71.7 cr males and 66.3 cr females living in India. India is the second-most populous country in the world behind China. It is now estimated that by 2027, India will most likely overtake China to become the most populous country in the world with 1.47 billion people. And by 2030, India will cross the 1.5 billion milestone. India's population will peak in 2059 with 1.65 billion people. Uttar Pradesh is the most populous state in India as well as the most populous country subdivision in the world. This densely populated state, located in the northern region of the Indian subcontinent, has over 200 million inhabitants, and the youth population (15-24 years) is near about 41 million.

India accounts for a meagre 2.4 percent of the world surface area, yet it supports and sustains a whopping 17.7 percent of the world population. Since the population of India is increasing at a slower rate than the world, its global share is decreasing. By 2100, 13.34% of the earth's population will be in India that is 4.42% less than the peak level of 17.76% in 2013.

The population growth rate for 2020 is projected at 0.99%, that is 117th highest among 235 countries/dependent territories. The population growth reached a peak in 1974 with an annual growth rate of 2.36%. India will add 13.6 million in 2020 that is near to the population of 74th ranked Somalia. The net increase of the population per year peaked in 1998 with almost 18.6 million.

According to the Census of India 2011, the population of India stood at 1,210,854,977. The percentage of decadal growth during 2001-2011 was 17.70%, 3.84% lower than the 1991-2001 period. The percentage of decadal growth during 2001-2011 has registered the sharpest decline since independence. The increased population during the decade 2001-2011 is approximately equal to the population of Pakistan, the sixth most populous country in the world.

The population of India, which at the turn of the twentieth century, was only around 238.4 million increased by more than five times in a period of 110 years to reach 1210 million in2011. Since independence the population of India has increased by 3.35 times. About 65% of the people of India live in rural areas and 35% live in urban areas. The proportion residing in urban areas has doubled in 2020 from 17% in 1950 and it is expected that half of the Indian population will be in urban areas by 2050. As of 2020, the share of the population in the age group 0-14 is 26.16 percent. The share of the working age population (15-65 years) is 67.27 percent. 6.57% of Indians are aged more than 60 years.

The fertility pattern of the developed regions of the world, consisting of Europe and North America, has caused the age structure of the population to shift upward, putting pressure on these nations to be dependent on the youthful nations from other parts of the world for labour supply. This decline in youth in the age group of 15-34, is seen as a prominent characteristic of developed regions, comprising Europe and North America since 1980. The regularity and efficiency of the census in India add rigor to the measurement of youth in India. The decennial enumeration through population census throws up a consistent estimate of the youth. As per India's Census 2011, the youth population (15-24 years) in India constitutes one-fifth (19.1 per cent) of India's total population.

India is expected to have a 34.33 per cent share of the youth in the total population by 2020. The share reached its maximum of 35.11 per cent in the year 2010, according to the annual growth rate. China, in contrast, is seen to have reached the highest share in the year 1990 at 38.28 per cent and is projected to have the share of the youth force shrinking to 27.62 per cent by the year 2020, a situation which Japan has experienced in around 2000.

It is observed that India has a relative advantage at present over other countries in terms of the distribution of the youth population. India is to remain younger longer than China and Indonesia, the two major countries other than India which determine the demographic features of Asia. These three countries together accounted for 68 per cent of the population of Asia in the year 2010, and the share of Asia was about 60 per cent of the world population.

The proportion of women in the youth age bracket, is generally lower, on account of the higher longevity of women compared to men. The difference on account of gender is seen to be higher in the developed region. In the case of India, gender differentials are less pronounced than in other countries. The shifts in the age distribution of the population to higher age groups result in a lower share for the age group 15-34 years, which in itself is an indication of increasing longevity. The widening differences on account of gender, characterizes such a situation, with the general population ageing and the female contribution is more, so leading to a still lower share of youth among women. Coincidently, in the case of India, the proximity of the share of the youth among men and women is indicative of the prevalence of healthy fertility levels in the general population.

Education is the key to overcoming development challenges in rural areas. There is not only a direct link between food security and the education of rural children, but it has also been shown that basic numeracy and literacy skills help improve farmers' livelihoods. Access to knowledge and information by the youth is crucial for addressing the main challenges they face in agriculture. In order for rural youth to shape agricultural policies affecting them directly, in terms of access to markets and finances, as well as green jobs and land, they need to receive appropriate information and education. While this is true in developed and developing countries alike, it is of particular concern in the latter, where young rural inhabitants may lack access to even the most rudimentary formal education and where educational institutions are often less developed. Formal primary and secondary education can provide young people with basic numeracy and literacy, managerial and business skills, and introduce youth to agriculture. Meanwhile, non-formal education (including vocational training and extension services) and tertiary agricultural education can offer youth more specific knowledge related to agriculture.

In developing countries, access to information and education is often worse in rural areas than in urban areas and this discrepancy is observable as early as primary school. In many rural areas of developing countries, children are malnourished and do not have the energy to attend a school or to easily absorb the information provided. During seasonal peaks in the agricultural cycle, there can be labour shortage and parents may see no other option than letting their children contribute to household and agricultural activities instead of attending school. The physical infrastructure of rural schools is often bad and classroom materials are sometimes lacking. Schools can be far away from rural communities making access difficult for rural children.

India is losing more than 2,000 farmers every single day and since 1991, the overall number of farmers has dropped by 15 million. This has several implications for the future of Indian agriculture and India's food security. Young farmers can play an important role in ensuring food security if they are encouraged to get involved in farming and the challenges they face are addressed. Over the past few years, rural youth have been shying away from agriculture and globally there is an increasing interest in finding ways of engaging youth in agriculture.

Social media is all about people. It is a way to build relationships, share information and connect with a diverse audience of people; you may never meet in real life. So interacting on social media, whether it is Twitter, Facebook, or Pinterest, allows you to develop a community and share your story in a way that was never possible before. The general public still has faith in farmers and ranchers, but some are still wary of modern farm practices. It is important that agriculture unites and has a chance to tell its side of the story.

Social media is one way to make the farmers' voices heard. It has the potential to reach farm families to educate them about children's health and safety. It offers advantages over traditional approaches because of the shorter time between creation and distribution and because of the greater reach and engagement possible. Recommendations are provided for how government agencies and the private sector can learn about and use social media to promote health and safety for children, as a supplement to traditional approaches. Social media usage continues to grow around the world, with global penetration rates now in excess of 50 per cent. "Facebook continues to dominate the global landscape, accounting for almost 2 billion users and is still adding around half a million new users every day or almost 6 new users every second," informed Simon Kemp on the 'We Are Social' blog. Facebook, India's most popular social network has last reported to have more than 241 million (17.95%) monthly active users While, Facebook and WhatsApp keep dominating the space, the growth of Google+ is interesting at a time when it is dismantling itself. Globally mobile is driving Facebook's growth and revenues.

The social media sites essentially pose a virtual representation of a user, called a profile. This profile often features the user's basic information, such as age, location and sex, as well as information regarding one's hobbies, such as favourite movies, musical artists and books. The most popular social networking services include My Space, Facebook, and Twitter. Many youth begin and end their day by checking posts on these social media sites. Decisions about how youth identify themselves, the feedback received on these decisions and how they view their own profile in comparison to others' profiles are potential factors in individual identity. The hyper-personal model for computer-mediated communication, for example, posts that youth engage in selective self-presentation online; moreover, the feedback from these presentations may in turn alter individuals' self-perceptions.

Also, the Internet makes it feasible for some youth to affiliate with other like-minded individuals online when such opportunities may not be possible in face-to-face interactions. That is, the youth can join "groups" reflecting aspects of their identity that they wish to explore or deepen and thereby foster a group identity. The youth can explore and expand their ideas and interests into new arenas through the Internet, e.g., communicating with others from more diverse backgrounds and expanding into new intellectual, political and social networks that create opportunities for transnational and global connections. Such connections can broaden as well as deepen self-identity, while at the same time enhance feelings of belongingness and affiliation. Participation of rural youth in all dimensions of the developmental process is essential in order to bring change in socio-economic status and in improving the quality of life for every individual of the community. Hence, much attention is needed to mould the personality of rural youth.

Therefore, the primary need of the day is to develop healthy and strong youth for the future development of the country. The rural youth comprising males and females are active partners in performing various agricultural activities, and as an "Agent of Change", they can help in the dissemination and subsequent adoption' process of modem agricultural techniques and practices among rural people.

3.2 Scaling Up Successful Models such as Contract Farming

Contract Farming: Contract farming refers to varied formal and informal agreements between producers and processors or buyers. It may include loose buying arrangements, simple purchase agreements and supervised production with input provision, with tied loans and risk coverage. In this system for the production and supply of agricultural/horticultural produce are under forward contracts between producers/suppliers and buyers. The essence of such an arrangement is the commitment of the producer/ seller to provide an agricultural commodity of a certain type, at a time and a price, and in quantity required by a known and committed buyer. Contract farming usually involves the following basic elements-pre-agreed price, quality, quantity or acreage (minimum/ maximum) and time. Contracts can range from oral deals to formal, registered written contracts.

Successful Contract Farming Models

There is a wide range of organizational structures that are comprised by the term "contract farming". The choice of the most appropriate one to use depends on the product, the resources of the company, the social and physical environments, the needs of the farmers and the local farming system. Some of the contract farming models practiced in India are presented below:

Private – Farmer Model

The poultry sector in Telangana made an impressive mark in promoting contract farming with private-farmer model as an effective institutional alternative for the promotion of broiler production. Suguna Poultry Farm Ltd. has emerged as one of the leading integrated broiler producers in the country, with contract farming. Contract poultry farming managed to free the small farmers from the worries of production and market planning of the poultry products.

Public-Private-Farmer Model

Maharashtra Department of Agriculture under 'tripartite public-private partnership' proposed contractual cotton cultivation programme in Akola, Amaravati, Buldhana, Wardha, Washim and Yavatmal districts in the state. The farming focus will be on cluster villages in these districts and the contract farming facilitation will be through the supply of credit-linked quality inputs to the farmers enrolled under the project which will be organised through the agricultural extension centers. The model to be followed will involve banking and insurance companies for credit input and loss guarantee and the choice of the cotton and their quality parameters would be decided in consultation with the Amaravati University, Maharashtra Industrial Development Corporation will facilitate by providing land for the textile sector in Amravati, the second biggest city after Nagpur in Vidarbha and the textile mills procure cotton in the State of Maharashtra from the State Co-Operative Cotton Growers Marketing Federation (MSCCGMF). The farming model would also receive incentive support from various State and Central Government schemes.

Private-Community Grower Group-Farmer Model

Ion Enviro undertakes contract farming with Community Grower Groups (CGGs) having large acreage, on a profit-sharing basis. Farmers are trained in-house in scientific organic farm management and certification. Community Grower Groups are promoted through Non-Governmental-Organisation (NGOs) or Self-Help Groups (SHGs) or registered associations. They follow fair trade practices wherein middlemen are eliminated, child labour is banned, men and women are given equal status, and transparency in trade is maintained. In the process, they bring to rural areas the best of organic processes and water management techniques, thereby educating and empowering farmers. Production is executed in accordance with protocol requirements as per Council Regulation (EEC) No. 2092/91 (The regulation provides the first Community rules for the production, labelling and control of agricultural products and foodstuffs produced organically, to ensure transparency at each stage of production and processing) standards. Written and documentary accounts are recorded to trace the origin, nature and quantities of raw materials procured and their usage (Ion exchange Enviro farms, 2005). The crops cultivated include Banana, Wheat, Cotton, Papaya, Pineapple, Basmati Rice, Mango, Soybean, Tur, Black Gram, Green Gram, Turmeric, Grapes, Bengal Gram, Groundnut, Sesame and Cashew.

Private Consortium - Farmer Model

Hindustan Lever Ltd (HLL), Rallis and ICICI jointly promote contract farming in wheat in Madhya Pradesh. Under the system, Rallis supplies agri-inputs and know-how, ICICI provide farm credit to the farmers and HLL buy-back the farm output. In this model, farmers benefit through the assured market for their produce in addition to timely, adequate and quality input supply including free technical know-how; HLL benefits through supply-chain efficiency; while

Rallis and ICICI benefit through assured clientele for their products and services. The model can be extended by including insurance firms, warehouses and manufacturers of equipment and machinery.

Industry - Research Institute- Farmer model

Pepsi Foods Ltd. entered India in 1989 by installing a tomato processing plant at Gahura in Hoshiarpur district of' Punjab to produce aseptically packed pastes and purees for the international market. Grower plants the company's crops on his land, and the company provides selected inputs like seeds/saplings, agricultural practices, and regular inspection of the crop and advisory services on crop management. The PepsiCo model of contract farming, in terms of' new options for farmers, productivity increases, and the introduction of modern technology, has been an unparalleled success. Another important factor in PepsiCo's success is the strategic partnership of the company with local bodies like the Punjab Agricultural University (PAU) and Punjab Agro Industries Corporation Ltd. PepsiCo is successfully emulating the above model in food grains, spices (chillies) and oilseeds (groundnut) and vegetable crops like potato. It also brings in the state of the art technology. At Sonipat, the company has an ISO 9002 and Hazard Analysis Critical Control Point (LIACCP) certified Rice Mill. Belgaum (Karnataka)-based Ugar Sugar Works Ltd., which established a successful backward linkage with farmers of Northern Karnataka for the supply of barley for its malt unit.

Grading House - SHG - Farmer Model

Appachi Cotton Company (ACC), the ginning and trading house in Pollachi under the name Integrated Cotton Cultivation (ICC), established backward and forward integration between the `grower' (farmer) and the `consumer' (textile units). The contract assured the farmers' easy availability of quality seeds, farm finance at an interest rate of 12% per annum, door delivery of unadulterated fertilisers and pesticides at discounted rates, expert advice and field supervision every alternate week, and a unique selling option through an MoU with ACC. The core principle of the formula lies in the formation of farmers' Self Help Groups (SHGs).

India's National Agriculture Policy envisages that "Private sector participation will be promoted through contract farming and land leasing arrangements to allow accelerated technology transfer, capital inflow and assured market for crop production, especially of oilseeds, cotton and horticultural crops". In India most farmers fall under small and marginal categories, with poor resources. Contract farming is generally defined as farming under an agreement between farmers and a sponsor (processing and/or marketing firm) for the production and supply of agricultural products under forward agreements, frequently at predetermined prices. Some examples of successful contract farming arrangements are cigarette factories, sugar mills and spinning mills. Contract Farming is gaining importance as a system that ensures the putting in place of backward linkages such as research, extension, financial services (including credit and insurance), input supply (seeds, fertilisers and pesticides) and forward linkages including storage, processing and marketing. It is a form of vertical integration within agricultural commodity chains so that the firm has greater control over the production process and final product. It also has the obvious advantage of guaranteeing a buyer to the producer – farmer, ensures and that the processor has a reliable supplier, with the government often offering its services to settle issues related to price and quality.

Advantages for Farmers

1. **Provision of Inputs and Production Services:** For ensuring proper crop husbandry practices in order to achieve projected yields in required qualities many contractual arrangements involve considerable production support in addition to the supply of basic inputs such as seed and fertilizer. Sponsors may also provide land preparation, field cultivation and harvesting as well as free training and extension.

2. **Access to Credit:** With the collapse or restructuring of many agricultural development banks, the majority of small holder producers experience difficulties in obtaining credit for production inputs. Contract farming usually allows farmers access to some form of a credit to finance production inputs. Arrangements can also be made with commercial banks or government agencies through crop-liens that are guaranteed by the sponsor, i.e., where the contract serves as collateral.

3. **Introduction of Appropriate Technology:** New production techniques are often necessary to increase productivity as well as to ensure that the commodity meets market demands. However, small marginal farmers are frequently reluctant to adopt new technologies because of the possible risks and costs involved. Private agribusiness usually offers technology more diligently than government agricultural extension services because it has a direct economic interest in improving farmers' production.

4. **Skill Transfer:** The skills the farmer learns through contract farming may include record keeping, efficient use of farm resources, improved methods of applying chemicals and fertilizers, knowledge of the importance of quality and the characteristics and demands of export markets. Farmers can gain experience in carrying out field activities following a strict timetable imposed by the extension service. In addition, spill over effects from contract farming activities could lead to investment in market infrastructure and human capital, thus improving the productivity of other farm activities.

Advantages for Sponsors

1. **Political Acceptability:** Contract farming, particularly when the farmer is not a tenant of the sponsor, is less likely to be subject to political criticism. It can be more politically expedient for a sponsor to involve smallholder farmers in production rather than to operate plantations. In recent years, many African governments have promoted contract farming as an alternative to private, corporate and state-owned plantations.

2. **Overcoming Land Constraints:** The majority of the world's plantations were established in the colonial era when land was relatively abundant, and the colonial powers had a little conscience about either simply annexing it or paying landowners the least compensation. At present, however, most large tracts of suitable land are either traditionally owned, costly to purchase or unavailable for commercial development. Contract farming, therefore, offers access to crop production farmland that would not otherwise be available to a company, with the additional advantage that it does not have to purchase it.

3. **Production Reliability and Shared Risk:** Working with contracted farmers facilitate sponsors to share the risk of production failure due to poor weather, disease, etc. The farmer takes the risk of loss of production while the company absorbs losses associated with reduced or non-existent output for the processing facility. Where production problems are widespread and for no fault of the farmers, sponsors will often defer repayment of production advances to the following season. Both estate and contract farming methods of obtaining raw materials are considerably more reliable than making purchases on the open market.

4. **Quality Consistency:** A steady market for fresh and processed agricultural produce requires reliable quality standards. Moreover, these markets are moving increasingly to a situation where the supplier must also conform to regulatory controls regarding production techniques, particularly the use of pesticides. Both estate and contracted crop production require close supervision to control and maintain product quality, especially when farmers are new with innovative harvesting and grading methods.

Disadvantages faced by the Farmers

1. **Possibility of Greater Risk:** Farmers entering into a new contract farming venture need to prepare themselves to assess the prospect of higher returns against the possibility of greater risk. Such risk is more expected when the agribusiness venture is introducing a new crop to the area. There may be production risks, particularly where prior field tests are inadequate, resulting in lower-than-expected yields for the farmers.

2. **Outdated Technologies and Crop Incongruity:** The introduction of a new crop to be grown under conditions meticulously controlled by the sponsor can cause disruption to the existing farming systems. Again, the introduction of sophisticated machines (e.g., for transplanting) may result in a loss of local employment and overcapitalization of the contracted farmer. Furthermore, in field activities such as transplanting and weed control, mechanical cultivation practices may produce fewer effective results than traditional ones.

3. **Manoeuvring in Quotas and Quality Specifications:** Incompetent management can lead production exceeding original targets. For example, failures of field staff to determine fields following transplanting can result in gross over planting. Sponsors may also have unrealistic expectations of the market for their product or the market may crumble unexpectedly owing to transport problems, civil unrest, changes in government policy or the arrival of competitors. In some situations, management may be tempted to manipulate quality standards in order to reduce purchases for honouring the contract. Such practices may cause sponsor-farmer confrontation, especially if there is no mechanism in place to settle disputes regarding grading irregularities.

4. **Corruption:** Difficulties arise when the staff responsible for issuing contracts and buying crops taking undue advantage of their position. Such practices result in the collapse of trust and communication between the contracted parties and soon undermine the contract. In a large contract, the sponsors may themselves be dishonest or corrupt. Governments have sometimes fallen victim to dubious or "fly-by-night" companies who have seen the opportunity for a quick profit. Therefore, farmers who make investments in production and primary processing facilities need to guard against such risks.

Disadvantages of Sponsors

1. **Limitation on Land Availability:** Farmers need to possess suitable cultivable land on which they are to cultivate contracted crops. But problems can occur when farmers have minimal or no security of tenure as there is a possibility of farmer - landlord disputes. Difficulties may also arise when sponsors lease land to farmers. Some contract farming ventures are dominated by customary land usage arrangements negotiated by landless farmers with traditional landowners. While such a situation allows the poorest cultivator to take part in contract farming ventures, discrete management measures need to be applied to ensure that landless farmers are not exploited by their landlords. Before signing a contract, the sponsor must ensure that access to land is secured, at least for the term of the agreement.

2. **Social and Cultural Constraints:** Promoting Contract Farming entails cultural, customary beliefs and religious issues. In communities where custom and tradition play an important role, difficulties may arise when innovative farming methods are introduced. Therefore, before introducing new cropping practices, sponsors must consider the social attitudes and the traditional farming procedures of the community and decide how a new crop can be introduced.

3. **Farmers' Disgruntlement:** Sometimes, situations may crop up which lead to farmers' discontent, e.g., biased buying, late payments, incompetent extension services, poor agronomic counsel, undependable transportation for crops, a mid-season change in pricing or management's impoliteness to farmers. They may aggravate the relationship between sponsors and farmers. If not readily addressed, such circumstances create antagonism towards the sponsors that may result in the farmers withdrawing from the project.

4 **Poor Quality Agro-input:** Sometimes, farmers are forced to use inputs supplied under contract for other than intended purposes such as for other cash and subsistence crops, or even to sell them. As a result, the contracted crop's yields may be reduced, and the quality affected. Improved monitoring by extension staff, farmer training and the issuing of realistic quantities of inputs can resolve such matters successfully. The majority of farmers will conform to the agreement when they have information that the contract has the advantages of technical inputs, cash advances and a guaranteed market. However, unless a project is very poorly managed, input diversion is usually an annoyance rather a serious problem.

5. **Sale of Crops by the Farmers beyond Contractual Agreement:** The sale of produce by farmers to a third party, outside the terms of a contract, can cause a major problem to the sponsors. However, extra-contractual sales are always possible when there is an alternative market. The outside buyers offered cash to farmers as opposed to the prolonged and difficult collection of payments negotiated through the cooperatives. Sometimes the sponsors may encourage extra-contractual practices as there are several companies working with the same crop (e.g., cotton in some southern African countries) and they could collaborate by establishing a register of contracted farmers.

Challenges of Contract Farming In India

Highly restrictive and regulated agricultural marketing system:

- The extant system in Indian markets is highly restrictive and regulated as only a few can enter into the market for purchasing produce from farmers as the government has a number of rules and regulations in place.

- Price setting not transparent - both producers and consumers are often cheated.

- Prices are fixed only by oral bidding and a company may cheat the farmer by purchasing at low prices as there is no bond between them. Similarly, if a farmer gets a higher price from another company then he may sell his produce to that company, rather than to the contracted one.

The monopoly of the State Governments in setting up markets:

- The APMCs set up by state governments are the only markets where the sale and purchase of produce are permitted by law.

- Mandi revenues not being deployed for infrastructure development is another weakness of the system.

Suggestions for Improvement of Contract Farming

- Adequate emphasis needs to be laid on infrastructure development.

- For a better understanding of the terms and conditions of the contract, the agreement should be in the regional language.

- Data should be maintained at the state and country levels to be made available for effective policy making.

- Success in developing contracting models or other forms of farm-firm linkages such as clusters that are effective for small holders will be a key to small holder participation.

- Contract farming is commodity specific and tends to promote monoculture. Thus land use planning is necessary together with the promotion of some incentives to contract farming firms.

3.3   Value Addition (Storage, Food Processing and Marketing)

Agriculture Value chains

The increased demand for safe, higher value and differentiated agricultural products has created opportunities for farmers and agribusiness entrepreneurs to transform commodities into products that are demanded by consumers. This change in food retailing has led to greater involvement of the private sector in agriculture and a focus on developing and improving agricultural value chains (AVCs) in terms of quality, productivity, efficiency, and depth.

The value chains are organized linkages between groups of producers, traders, processors, and service providers (including nongovernment organizations) that join together to improve productivity and the value added from their activities. In a well-managed value chain, the value of the end-product is often greater than the sum of individual value additions. By joining together, the members in a value chain increase competitiveness and are better able to maintain competitiveness through innovation. The limitations of each single member in the chain are overcome by establishing collaborations and governance rules aimed at producing higher value. The main advantages to commercial stakeholders from being part of an effective value chain comprise being able to reduce the cost of doing business, increase revenues, increase bargaining power, improve access to technology, information, capital and, by doing so, innovate production and marketing processes to gain higher value and provide higher quality to customers.

The expression "farm-to-fork" is often used to describe food to elaborate VCs. This means that a food product moves from upstream in the chain, where farmers grow and harvest it, towards the market – through intermediaries including producer organizations, processors, transporters, wholesalers and retailers – and on to the downstream level of consumers. A value chain system can be defined as a strategic partnership between inter-dependent businesses that comes together to progressively create value for the final consumer, resulting in a collective competition advantage.

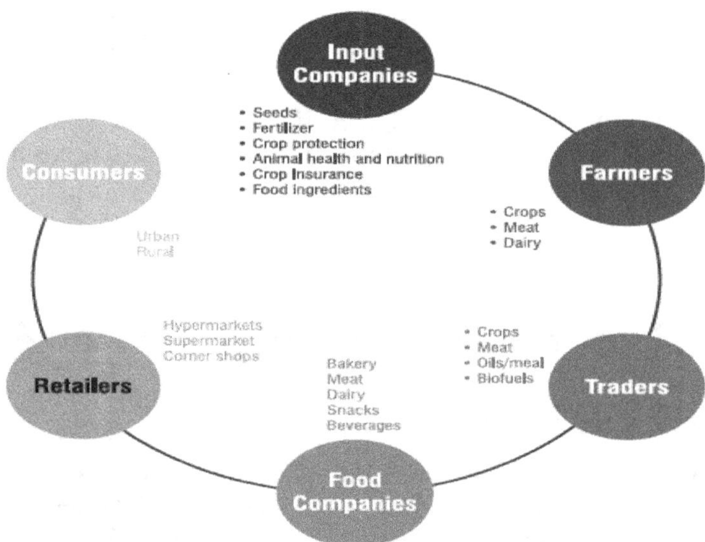

Figure 3.2 Agriculture Value Chain system

The first and foremost thought that comes to my mind is to "Make the country free of hunger".

Capturing the value created along the agriculture chain from pre-sowing to food harvest or in other words capturing value created from "Farm to Fork".

The agriculture value chain needs to prevent colossal wastage of agriculture produce "billions of dollars" loss to the economy, ensure that share of farmer in consumers wallet can increase, which is merely 25-30% as compared to the western market where it goes as high as 50-75 per cent. Facilitate demand for Minimum Support Price (MSP) regime which will automatically go away and farmer can expect a rightful price for the produce, ensure optimal management of natural resources and mother earth which is being abused, make that India becomes "Global Hub for Food Industry". Food wastage in India is among the biggest food wasters in the world, reportedly wasting an estimated ₹ 900,000 million worth of fruits, vegetables and grains every year and year on year. India wastes wheat equivalent Australia grows in a year more fruits and vegetables are wasted than that consumed by the United Kingdom in a year.

Current Scenario of Indian Food Market

India houses 18% of the world's human and 15% of the livestock population, while subsisting on just 2.2% of its geographical area, 4.2% of freshwater resources, 1% of forest and 0.5% of pastureland. On top of these are the challenges of climate change, shrinking holding size (more than 80% of the farmers are small and marginal) and nearly 52% of the agricultural land being un-irrigated.

Even in the face of such adversities, our farmers — by marrying their traditional knowledge with the adoption of technology, high-yielding and climate-resilient varieties/hybrids and modern inputs — have made India largely self-sufficient through record production of 285 million tonnes of food grains and 311 million tonnes of fruits and vegetables.

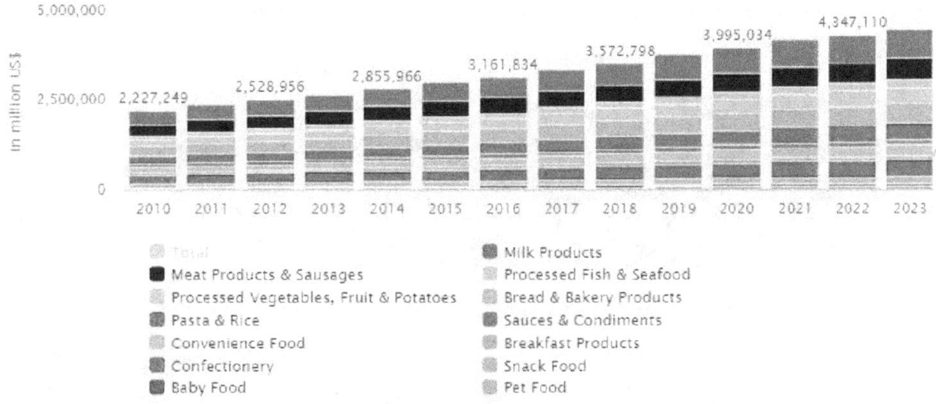

Figure 3.3 Current status of the Indian Food Market

*Source:*www.statistia.com

With a view to giving a strong push to the setting up of food processing infrastructure, the Pradhan Mantri Kisan Sampada Yojana was launched in May 2017, with a planned outlay of ₹ 6,000 crore. Key focus areas include the development of Mega Food Parks, integrated cold chain and value addition infrastructure, agro-processing clusters, food preservation and processing capacities, food testing laboratories and backward and forward linkages. Between 2014 and 2018, 13 Mega Food Parks have become operational, benefitting 20,725 farmers and generating employment for 334,854 people, and 27 more are under implementation. During the same period, 85 cold chain projects have been operationalized with capacity addition for cold storage recorded at 2.76 lakh.

Even as the share of manufacturing and services in the Indian economy grows, agriculture still accounts for around 49% share in employment 'Promotion of value addition through food processing' is a major plank of the seven-point strategy outlined by the Hon'ble Prime Minister to double the income of farmers.

Rapid advancements in food processing infrastructure will be of immense benefit to Indian farmers in the coming years – it will help them prevent wastage, promote value addition and ensure that they get remunerative prices for their produce and tide over the vagaries of demand and supply. Moreover, improvements in food quality and value addition will open many more avenues for producers in domestic as well as in export markets.

Agriculture Storage Infrastructure

The post-production stages of the agricultural value chain need to be effectively supported by supporting infrastructure. India faces an impeding challenge of large post-harvest losses. As per estimates by the Associated Chambers of Commerce of India, post-harvest losses (PHL) stand at INR 926 billion (US$14.33 billion). It is estimated that 3.9-6% of cereals, 4.3- 6.1% of pulses,

2.8- 10.1% of oilseeds, 5.8-18.1% of fruits, and 6.9-13% of vegetables suffer from losses during harvest and post-harvest stages. One of the major factors for these losses is the inefficient post-harvest and storage infrastructure, further intensifying the need to undertake large-scale improvements.

In India, three main agencies in the public sector build large-scale storage and warehousing capacity, viz., Food Corporation of India (FCI), Central Warehousing Corporation (CWC) and State Warehousing Corporations (SWCs). The main functions of the CWC and SWCs are to acquire and build warehouses at suitable places and to operate them for storage of agricultural produce, fertilisers, and certain other items, including industrial goods.

The majority of small and marginal farmers lack access to the economies of scale. This may be tackled by the creation of collection hubs at the village level to encourage farmers to use the facilities. Further, in the cold chain, the first mile capacity to facilitate market connectivity of short life horticultural produce is a challenge. Therefore, building aggregation units (i.e., modern pack-houses and pooling points) at the village level with transport links would help strengthen the chain.

It is necessary to introduce a cluster-focused approach for the development of cold storage to enable crop specific support in particular regions. The scheme for cold storage could include a separate action plan to focus on the development of these clusters with the State as the responsible authority. It is necessary to develop modern infrastructure with facilities such as scientific warehousing, GPS tracking, real-time monitoring. Public-private partnership (PPP) can be a useful instrument in developing infrastructure. Provision of loans at the same interest rate and subvention provision as short-term crop loans will encourage farmers to store in the warehouses and get necessary credit facilities at optimum rates.

Food Processing

Food processing is an important sector as it provides a strong link between agriculture and the end– consumer. Food processing is a set of methods and techniques used to transform raw agricultural produce into a form that can be consumed directly. It involves any type of value addition to agriculture or horticulture produce that enhances the shelf–life of the food product.

The food processing industry is made up of two kinds of processing:

1. **Primary processing:** It includes conversion of raw farm output to intermediate commodity consumables with activities like shelling, hulling, milling, polishing, crushing, packing, etc. It is required for certain farm products only — cereals, pulses and oilseeds. Examples of primary processed food sold to the end–consumer include packaged fruits and vegetables.

2. Value-added processing includes conversion of raw or intermediate farm output to value-added products with activities like flour milling, baking, fortification, refining, etc. Examples of value-added processed food sold to end– consumers include juices, jams, pickles, squashes, concentrate, ghee, paneer, cheese, butter, ethnic Indian products, branded edible oil, breads, biscuits, snack foods, pasta– based foods, processed meat, poultry, marine products confectionery and chocolates, beer, spirits, wine, aerated, and malted beverages.

India is one of the key food producers in the world, with the second-largest arable land area. It is the largest producer of milk, pulses, sugarcane and tea in the world and the second-largest producer of wheat, rice, fruits and vegetables. India's Food Processing industry is one of the largest industries in the country - it is ranked fifth in terms of production, consumption, export and expected growth. The Indian food processing industry accounts for 32 per cent of the country's total food market, one of the largest industries in India and is ranked fifth in terms of production, consumption, export and expected growth. It contributes around 8.80 and 8.39 per cent of Gross Value Added (GVA) in Manufacturing and Agriculture respectively.

Food Processing includes the process under which any raw product of agriculture, dairy, animal husbandry, meat, poultry or fishing is transformed through a process (involving employees, power, machines or money) in such a way that its original physical properties undergo a change and the transformed product has commercial value and is suitable for human and animal consumption. It also includes the process of value addition to produce products through methods such as preservation, the addition of food additives, drying, etc. with a view to preserve food substances in an effective manner, enhance their shelf life and quality.

The food processing sector needs to develop linkages in agricultural value chains in India to make it more remunerative for farmers. Food processing forms an integral part of the forward linkages assisting in the reduction of food wastage and increasing the value of produce through value addition. India, with a large consumer base of 1.3 billion and as a major agricultural producer, could be a major food-processing hub. The. The industry engages approximately 1.77m people in around 39,319 registered units with fixed capital of US$29.2 billion and aggregate output of around US$144.6 billion.

Food Processing Industry Importance

The Food Processing Industry (FPI) is of enormous importance as it provides vital linkages and synergies that it promotes between the two pillars of the economy, i.e., agriculture and industry.

- **Employment Generation:** It provides direct and indirect employment opportunities, because it acts as a bridge between Agriculture and Manufacturing.
- **Doubling of farmers' income:** With the rise in demand for agri-products there will be a commensurate rise in the price paid to the farmer, thereby increasing the income.
- **Reduce malnutrition:** Processed foods when fortified with vitamins and minerals can reduce the nutritional gap in the population.
- **Reduce food wastage:** UN estimates that 40% of production is wasted. Similarly, NITI Aayog estimated annual post-harvest losses of close to ₹ 90,000 crore. With a greater thrust on proper sorting and grading close to the farm gate, and diverting extra produce to FPI, this wastage could also be reduced, leading to better price realisation for farmers.
- **Boosts Trade and Earns Foreign exchange:** It is an important source of foreign exchange. For e.g., Indian Basmati rice is in great demand in Middle Eastern countries.
- **Curbing Migration:** Food processing being a labour-intensive industry will provide localized employment opportunities and thus will reduce the push factor in source regions of migration.
- **Curbing Food Inflation:** Processing increases the shelf life of the food, thus keeping supplies in tune with the demand thereby controlling food-inflation. For e.g., Frozen Safal peas are available throughout the year.
- **Crop-diversification:** Food processing will require different types of inputs, thus creating an incentive for the farmer to grow and diversify crops.
- Preserve the nutritive quality of food and prolongs the shelf life by preventing them from spoilage due to microbes and other spoilage agents,
- Enhances the quality and taste of food thereby bringing more choices in the food basket
- **Enhances consumer choices:** Today, food processing allows food from other parts of the world to be transported to our local market and vice versa.

Prime Minister Kisan Sampada Yojana (PMKSY)

PMKSY is a comprehensive package, which will result in the creation of modern infrastructure with efficient supply chain management from farm gate to retail outlet. The emphasis is on the creation of infrastructure facilities which include

1. Mega food parks;
2. Integrated cold chain and value addition infrastructure;
3. Creation/ Expansion of food processing and preservation capacities;

4. Infrastructure for agro-processing clusters;
5. Creation of backward and forward linkages;
6. Food safety and quality assurance infrastructure;
7. Human resources and institutions.

Although India has allowed 100% FDI in food processing, the process is quite tedious and complex and needs to be simplified. Food processing is comprised of primary, secondary and tertiary processing. There needs to be incentivisation for investments in secondary and tertiary processing needs more promotion at all scales. It is necessary to overcome fragmented supply chains through the promotion of initiatives such as contract and corporate farming, the establishment of FPOs to directly link with food processors to ensure quality and consistency in raw material supply and optimum remuneration to farmers.

Developing a national level food processing policy is necessary to address key areas of reform in the sector such as private sector participation, promotion of export, infrastructure and institutional strengthening. While some states such as Andhra Pradesh, Jharkhand, Odisha and Tamil Nadu have come up with State Food processing policies, and a national level policy is needed, with various reforms charted out for the food processing sector. Separate assistance for skill development in the food processing sector is also required.

Agriculture Marketing

Indian agriculture marketing system is characterised by the presence of multiple intermediaries often leading to challenges such as non-optimum price realisation, information and transaction leakages. Over the past years, the Indian agricultural marketing system has evolved, considering the changes in the agriculture sector. However, it is necessary to implement these initiatives on a large scale for the benefit of farmers. The status of marketing reforms in India is illustrated below, with the progress of seven key marketing reforms across the states. The implementation of reforms is non-uniform and needs to be incorporated by all State agriculture marketing departments/ boards.

National Agricultural Market (NAM)

The Government introduced a Central Sector Scheme for the promotion of a National Agricultural Market (NAM) to transform the agricultural marketing environment. A unified market can be best realised through a pan-India electronic platform, which can facilitate the participation of buyers and sellers from all over. The e-NAM network was inaugurated on 14 April 2016 and until date, 585 markets, 16 States and 2 Union Territories have been integrated with e-NAM.

The performance of Indian agriculture is directly linked to the access of farmers to input and output markets. There has been a considerable process in this direction, but agricultural markets need to be modernized for inclusiveness and efficiency. The system of agricultural marketing should be strengthened and integrated to meet the rising demand of consumers on the one hand and, on the other, farmers should be able to realize a higher price.

Recent reforms such as the Agricultural Produce Marketing Committee (APMC) Act, contract farming and eNAM can be cited as measures to improve market integration, infrastructure and technology use. However, state participation and compliance are highly erratic. Even existing market establishments prove to be inadequate, thus hampering access to the market. Against the recommendation of a regulated market in each 80 sq. km, the present scenario throws out a picture of an average coverage of 490 sq. km/market. The country has just one-sixth of the recommended strength. Furthermore, farmer's awareness of price support measures does not seem encouraging. The situation assessment survey of the National Sample Survey Office (NSSO) points out that just 30% and 40% of farmers reporting the scale of paddy and wheat, respectively, are aware of the existence of the Minimum Support Price (MSP) instrument.

Awareness about procurement agencies is still less convincing. Improving market access through new establishments, infrastructure and technology upgrades, participation compliance and transparency are all immediate measures that are needed in the agricultural marketing sector. There are some parts of the country, like eastern India, where much investment is needed in rural infrastructure along with greater emphasis on agricultural marketing.

Setting up agriculture commodity markets in accessible locations can help increase farmer participation in formal marketing systems. Primary agriculture markets operating in approximately every five km radius would increase the ease of selling and make it more economically viable for farmers.

e-NAM operation requires competency in operational procedures such as online registration, payment, data management. Handholding support to the e-NAM mandis for the first three to four years and module-based structured training regularly can bring efficiency among the staff. It is necessary to deploy efficient staff for specific job roles of e-NAM for longer periods. e-NAM enables buyers to participate in inter-mandi and inter-State trade of commodities. However, after the trade, buyers from distant locations have to make all logistical arrangements to move the commodity. Organised logistics services by registered vendors of e-NAM mandi can ease the transaction process and encourage more inter-mandi and inter-State trade.

Linking farmers with current market information and trends through ICT based channels can help them minimise informational asymmetries and increase their negotiation power in the market channel. This could help farmers in decision making to select the time and place of selling their commodities.

It is necessary to provide more marketing facilities by increasing the efficiency of existing markets, creating new markets, providing agriculture marketing extension services. A conducive environment and required support is required to promote private markets, direct marketing place and contract farming.

3.4 Precision Agriculture and Smart Farming

Precision agriculture can be defined as "the application of modern information technologies to provide, process and analyse multi-source data of high spatial and temporal resolution for decision making and operations in the management of crop production".

Precision Farming can be defined as managing crop production inputs (seed, fertilizer, lime, pesticides, etc.) on a site-specific basis to increase profits, reduce waste and maintain environmental quality.

The world needs a second "green revolution". Precision farming, smart farming and digital farming are the three approaches to agriculture that will enable this necessary reform.

Precision Agriculture

Precision farming is a management strategy. Its development can be attributed to two major trends: on the one hand, it uses big data, advanced data analysis and robotics; on the other it relies on aerial images, sensors and sophisticated weather forecasts. The focal point of this type of agriculture is to improve production efficiency.

Often, the yield of a field is not homogeneous. There are areas where a plot of land is more productive and others where it is less so. The causes are different, measurable. Understanding and calculating them in order to solve the situation is the main objective of precision farming.

For example: a farm operating under a precision farming regime will never apply the same fertilizer to all sections of its land. On the contrary, it will analyse the soil composition and adapt the fertilization strategy to the individual characteristics of each piece of land.

Smart Farming

Smart farming is the application of information technology for the optimization of agricultural systems. Unlike precision farming, where the focus is on measuring and resolving inconsistencies within the same field, this approach focuses on how to access and apply data. In practice, it studies the best way to collect and use information. According to smart farming, a farmer should always have a smartphone or tablet at his fingertips, so that he can access live data on soil, plant, climate, weather and resource conditions. The purpose of this implementation is to enable the farmer to make informed decisions, in real-time, based on concrete data.

Digital Farming

Digital farming is a complementary step to smart farming and its main responsibility lies in giving value to the data collected. This approach integrates the concepts of precision farming and smart farming and can be defined as the consistent application of both methods. Digital agriculture takes precision farming technologies and refines them, combining them with data management tools. The objective of this process is to use all available information to enable the automation of certain processes applied in agriculture. A digital farm must use machines that can send and receive information through sensors, facilitate automated operations and assist people as much as possible.

In the near future, in order to increase agricultural production while also protecting the environment, it is necessary to increase the sustainability of the agricultural model through innovation. The use of new technologies results in two closely interrelated economic benefits: increasing productivity and reducing the environmental impact of an agricultural system. The need to improve productivity, competitiveness and environmental performance do not only concern the economic sphere.

In order that increases in production and enhancement in yield prove equal to the future challenges of food and nutritional insecurity, the adoption of new technologies is a must. It is also essential that such technologies are eco-friendly. Since long, it has been recognized that crops and soils are not uniform even within a given field. Over the last decade, methods have been developed to utilize modern electronics to respond to field variability. Such methods are known as spatially variable crop production, a geographic positioning system (GPS)-based agriculture, site-specific and precision farming (precision agriculture). The term 'spatially variable crop production' appears the most-appropriate description.

Precision agriculture uses recent developments such as sensors, green-houses and protected agriculture structures. This technology can be meaningfully deployed for hot and extremely dry regions where water is scarce, the soils saline, temperatures high and rainfall low. Even in developing countries, the availability of labour for agricultural activities is going to be in short supply in the future. The time has now come to exploit all the modern tools available by bringing information technology and agricultural science together for improved and environmentally sustainable crop production. Precision agriculture is an integrated crop management system that attempts to match the type and quantity of inputs with the actual crop needs for small areas within a farm field. This attempt is not new, but new technologies now available allow the concept of Precision agriculture to be realized in a practical production setting.

The potential of Precision Farming for economic and environmental benefits can be realized through reduced use of water, fertilizers, herbicides and pesticides besides the farm equipment. Instead of managing an entire field based upon hypothetical average conditions, which may not exist at any point in the field, a precision farming approach recognizes site-specific differences within fields and adjusts management actions accordingly. Precision Agriculture offers the potential to automate and simplify the collection and analysis of information and management decisions to be made and quickly implemented in small areas within larger fields.

Effective use of inputs means greater crop yield and/or quality, without polluting the environment. Precise determination of the benefit cost of Precision Agriculture management, however, remains difficult.

The concept of "doing the right thing in the right place at the right time" has however, a strong intuitive appeal. Ultimately, the success of Precision Agriculture depends largely on how well and how quickly the knowledge needed to guide this new technology can be required. Tools are also available now to apply chemicals, fertilizers, tillage, and seed differentially to a field and collect the yield or plant biomass by position across the field. Remote sensing technology allows observation of variations within a field throughout the growing season, as opposed to impose management changes. Monitoring equipment is also available for capturing the surface water and groundwater samples needed to quantify the environmental impact through surface runoff or leaching. Technology also exists to capture the volatilization of nitrogen or pesticides from the field into the atmosphere from modified practices.

Advantages of Precision Farming

1. It enhances agricultural productivity and prevents soil degradation in cultivable land resulting in sustained agricultural development.

2. It reduces excessive chemical usage in crop production.

3. Water resources will be utilized efficiently under precision farming.

4. GPS allows agricultural fields to be surveyed with ease. Moreover, yield and soil characteristics can also be mapped.

5. Dissemination of information about agricultural practices to improve the quality, quantity and reduced cost of production in agricultural crops.

6. It minimizes the risk to the environment particularly with respect to nitrate leaching and groundwater contamination by means of the optimization of agro-chemical products.

7. Non-uniform fields can be sub-divided into smaller plots based on their unique requirements.

8. It provides opportunities for better resource management and hence reduce the wastage of resources

Disadvantages of Precision Farming

High capital costs may discourage farmers to not adopt this method of farming. Precision agriculture techniques are still under development and require expert advice before actual implementation. It may take several years before the actual collection of sufficient data to fully implement the system. It is an extremely difficult task particularly the collection and analysis of data.

How could India benefit from precision farming?

1. Refinement and wider application of precision agriculture technologies in India can help in reducing production costs, increasing productivity and better utilization of natural resources.

2. It has the ability to revolutionize modern farm management in India through improvement in profitability, productivity, sustainability, crop quality, environmental protection, on-farm quality of life, food safety and rural economic development.

3. Site-specific application of irrigation in wheat of Punjab and Haryana, pesticides in cotton and fertilizers applications in the oil palm plantation in South India, and coffee and tea garden of eastern India can highly reduce production costs and also reduce environmental loading of chemicals.

4. It can increase the efficiency of irrigation efficiency when water resources are low. Farmers can use forecast and mitigate problems like water stress, nutrient deficiency, and pests/diseases.

5. It also increases opportunities for skilled employment in the agriculture sector and also provides new tools for evaluating multifunctional aspects including non-market functions.

6. It has an essential role in the monitoring of greenhouse conditions in agricultural fields.

Challenges to Adopting Precision Farming in India

- The adoption of precision farming in India is yet in the nascent stage on account of its unique pattern of land holdings, poor infrastructure, and lack of farmers' inclination to take the risk, social and economic conditions and demographic conditions.

- The small size of landholdings in most of the Indian agriculture limits economic gains from currently available precision farming technology.

Precision Agriculture Technologies (PATs)

Precision Agriculture Technologies (PATs) are one of the most efficient tools to improve sustainability and productivity in farming. PATs offer solutions to produce more with less and enhance food security and safety. Practically, PATs provide farmers with extra sensors which give them more information on how to manage natural variations such as weather conditions, pests, insects and fungal infestations. Essentially, they help in:

- Preventing ground water pollution by optimizing manure and chemical spraying
- Reducing fresh water withdrawals with precision irrigation
- Limiting crop damages by responding rapidly and effectively to pest and fungal infestation
- Allowing new types of poly-culture (critical to stimulate biodiversity, noticeably for pollinators)

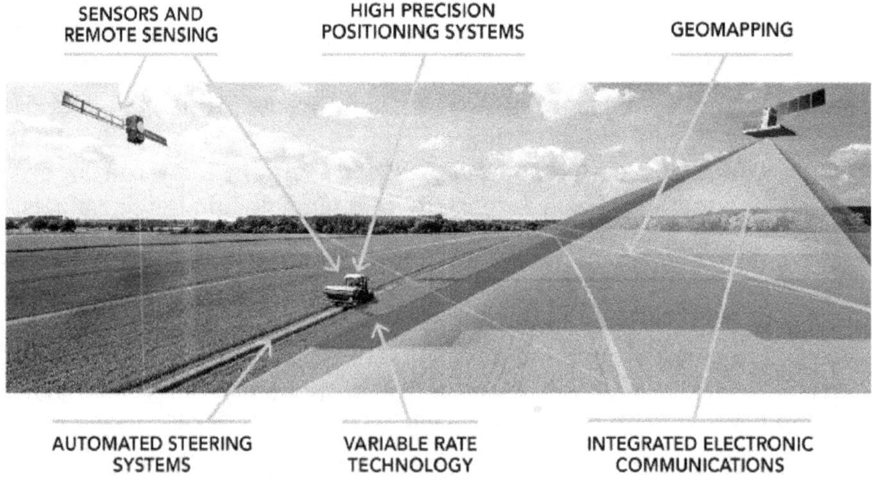

Some precision agriculture diagnostic technologies are already highly affordable and thus available to smaller farms thanks to smart phones or tablets and their applications. Such applications can directly signal a problem on the field or connect to an online service for further probing. Other fundamental PATs are less available to smaller farms and should therefore be promoted by a policy. Such technologies can be divided into three categories:

1. Guidance Systems
2. Variable Rate Applications (VRT) and Nutrient Sensing
3. Precision Livestock Farming (PLF) Technologies

Pest-control is, in a certain way, even more critical for small-sized farms than for larger farms. Due to the way pests disseminate – from a spot to the entire field, small farms have a shorter time to react and they face the risk to lose their entire crops, while bigger farms have somewhat more time to react and are also better equipped to limit the damages to a certain percentage of their crops.

1. **Guidance Systems:** Guidance systems form the generic backbone technology for Precision Agriculture. They can be used by all kinds of equipment (e.g., tractors, combine-harvesters, sprayers, planters, etc.) and as part of a broad range of different agricultural applications. Guidance systems focus on precise positioning and movement of the machine with the support of a Global Navigation Satellite System (GNSS).

Guidance Systems enable:

- Automatic steering
- Precise machine movement between plant rows
- Precision drilling and sowing
- Precision spraying
- Mechanical weeding
- Field digitalisation

The most tangible benefits of Guidance technologies are:

- Minimising overlapping by increasing pass-to-pass efficiency leading to lower fuel consumption (up to 10% less fuel consumption)
- Reduction of all agricultural inputs (seeds, herbicides, pesticides, fertilizers, etc.)

It is evident that guidance systems have a positive impact on:

- Reducing farm-related GHG emissions and other pollutants,
- Boosting the profitability and productivity of farm holdings, as fewer inputs and less work are needed for maintaining, or even increasing, yields levels.

2. **Variable Rate Technologies (VRT):** Variable rate technologies (VRT) or variable rate applications (VRA) have features that allow variations in the rate of applications to conform to the specific needs of the plants, which depends for instance on the yield variability, within the same field.

   VRA technologies are mostly used for spraying:
   - Water
   - Pesticides
   - Herbicides

- Fungicides
- Inorganic and manure fertilisers

Geo-Referenced Fields: Once the field is Geo-referenced, a prescription can be used for variable rate application:

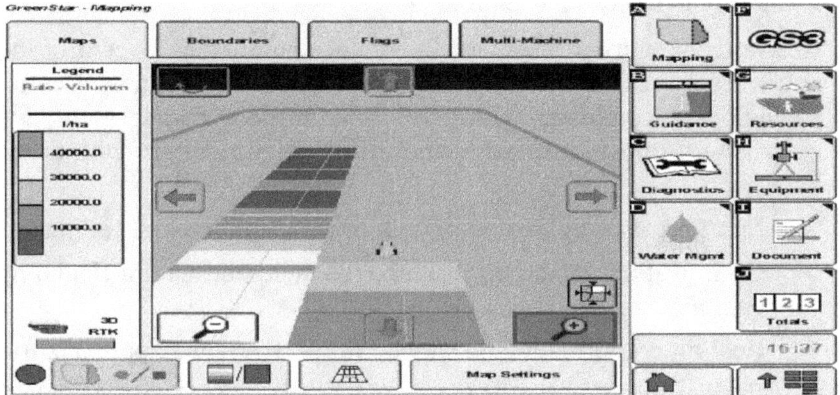

Selective spraying is possible if a disease sensor and a controller are available with the variable rate application system, as shown in the graph below:

Selective Spraying for disease control

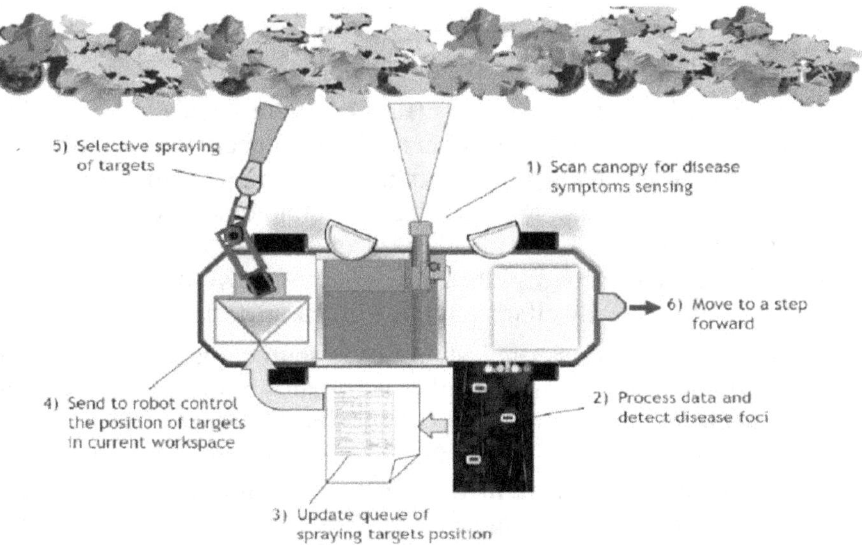

Figure 3.4 Fungicide reduction 20-30% (max 80%)

Nutrient Sensing Technologies: Successfully using manure as a fertilizer requires an assessment of the available nutrients in manure, calculating the appropriate rate to provide the needed nutrients to the crop and applying the manure uniformly across the field at the target rate. Efficient use of manure as a fertilizer is complicated by the imbalance of nutrients in manure, variability in many sources of manure.

Now-a-days, nutrient sensing technologies automatically control the desired nutrient application rates on the go more accurately them ever before. These technologies combined with VRT and nutrient management definitively contribute to more sustainable and profitable agriculture with the following benefits:

- More precise application of both organic and mineral fertilizer.
- Optimization of nutrient balance (by field and within fields) during the complete growing season.
- Real time information on supplied-, received- and applied nutrients.
- Facilitating the application of manure based on actual Nitrogen, Phosphorus, Potassium (NPK) values (kg/ha) i/o volume (m^3/ha).
- Fully compensating for variability of manure nutrient ingredient contents.
- Easier and better agronomic decision support and documentation.
- Maximizing crop yield potential with environmental protection.
- Facilitating more sustainable crop production and soil fertility management.

3. **Precision Livestock Farming (PLF) Technologies:** PLF technologies include a wide variety of machines, farm management systems and other devices used for livestock farming. They range from advanced feeding systems to cleaning and milking robots. Technologies like these are available and beneficial for all types and sizes of farms. In the earlier years these technologies have progressed rapidly and now have proven benefits for the environment, animal welfare and farmers' competitiveness.

Smart Agriculture: Farm Enterprise Resource Planning (ERP) refers to software and systems used to plan and manage all the core supply chain, manufacturing, services, financial and other processes of an organization. It is the most comprehensive and successful flagship software solution being used globally for farm, farmer, procurement, processing, supply chain and financial data management and analysis. It has helped many leading companies achieve profitable and sustainable agribusiness. Smart agriculture devices can help in acquiring pertinent farm data and provide actionable insights to the farmers. Internet of Things (IoT)can help in enhancing farm management efforts leading to better agriculture outcomes.

Meeting the rapidly rising demand for food, especially in the context of the increasing frequency and intensity of extreme weather events is proving to be a great challenge for the country. Growing incidents of pest infestations and crop disease are causing huge losses to farmers. Labour shortage and livestock illness are further taking a toll on farm management practices, affecting the livelihoods of vulnerable farmers and threatening the nutrition security of millions. The impact of smart devices has been invariably positive since the beginning. IoT-driven agriculture is significantly boosting farm management efforts and enhancing the quality of crop yield. The use of smart sensors, drones and other devices can encourage the adoption of precision farming and enable data-driven decision making.

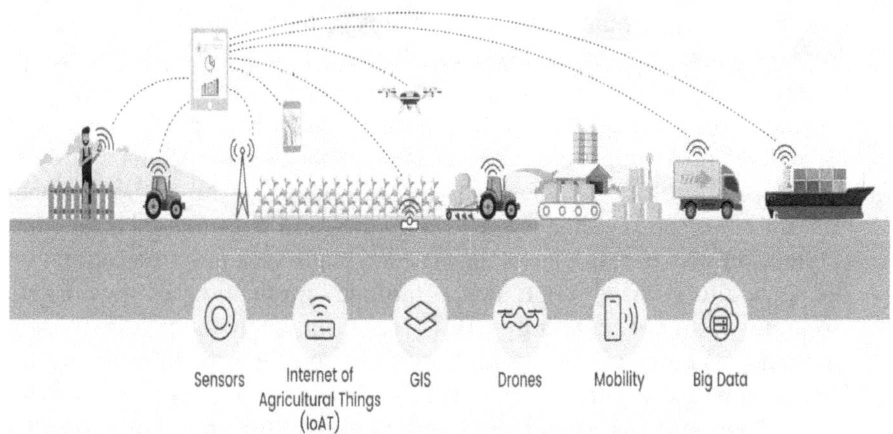

Farm ERP and FarmGYAN are helpful to farm managers and agribusinesses around the globe to collect, integrate and analyse huge amounts of data to support their business decisions. FarmGYAN intends to provide science-based intelligent advisory services to farm operators. These services are based on the Smart Agriculture based framework. This advisory is customized for a particular plot and works at the macro level. Smart Agriculture framework combines the power of Farm ERP with Big Data, Internet of Agricultural Things (IoAT), Sensors, Robotics, Geographic Information Systems, Unmanned Aerial Vehicle (UAV) / Drones, Precision agriculture, satellite data, real-time cloud data sources and predictive analysis.

Smart Farming: Smart Farming is an emerging concept that refers to managing farms using modern Information and Communication Technologies to increase the quantity and quality of products while optimizing the human labour required.

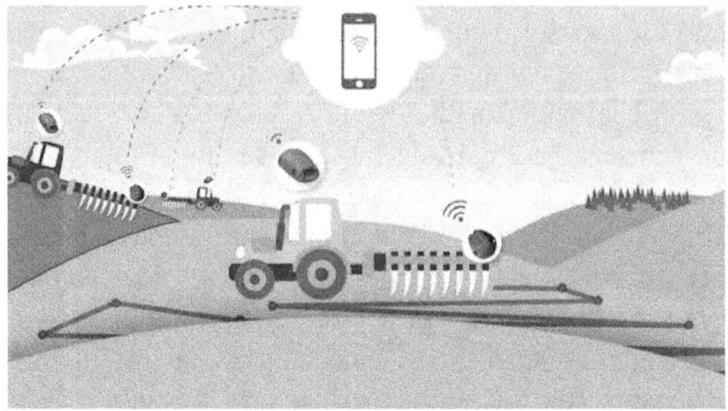

The induction of modern Information and Communication Technology (ICT) in agriculture, leading to what can be called a third Green Revolution is referred to as Smart Farming. It helps reduce overall costs and improves the quality and quantity of products, increasing control over production leads to improved cost management and waste reduction. The ability to trace anomalies in crop growth or livestock health, for example, helps eliminate the risk of dropping yields. Also, automation boosts efficiency. With smart devices, multiple processes can be activated at the same instance, and automated services enhance product quality and volume by better controlling production processes. The plant breeding and genetics revolutions takeover the agriculture-based upon the combined application of Information and Communication Technology (ICT) solution such as precision tools, the Internet of Things (IoT), sensors and actuators, geo-positioning system, Big Data, Unmanned Aerial Vehicles (UAVs, drones), robotics, etc. However, Smart Farming has a real potential to distribute an extra productive and sustainable agriculture, based on a more precise and resource efficient approach.

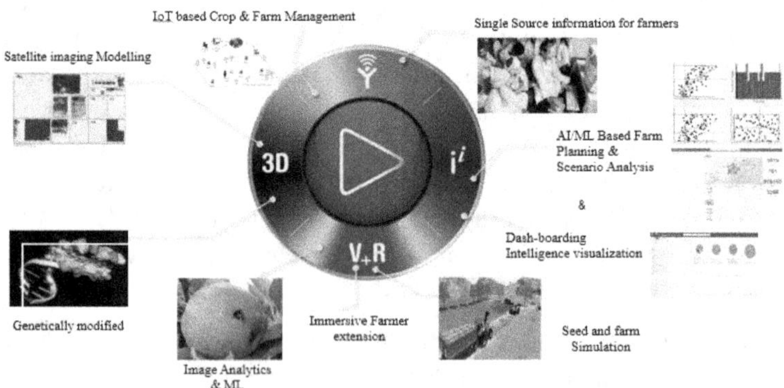

Figure 3.5 3D Experience Technology – New Generation Agriculture Technologies

Digital Soil Information on Mobile Appliances for Farmers

- Updated Soil Information
- To Improve Yield/Predictability
- Yield and Profit maximization incorporating spatial variability and nutrient response.
- The collaboration of farmers with agronomists on decision making. (Fertilizer application, data sheets)
- Crop performances and weed infestations
- UAV control and configurations
- Manage harvesting vehicles
- Connect and resolve support issues to farmers
- Determining Nitrogen uptake in crops, chlorophyll content, etc.
- Location specific rainfall measurements and predictions
- Careers in agriculture
- Build maps of farms and design your dream field

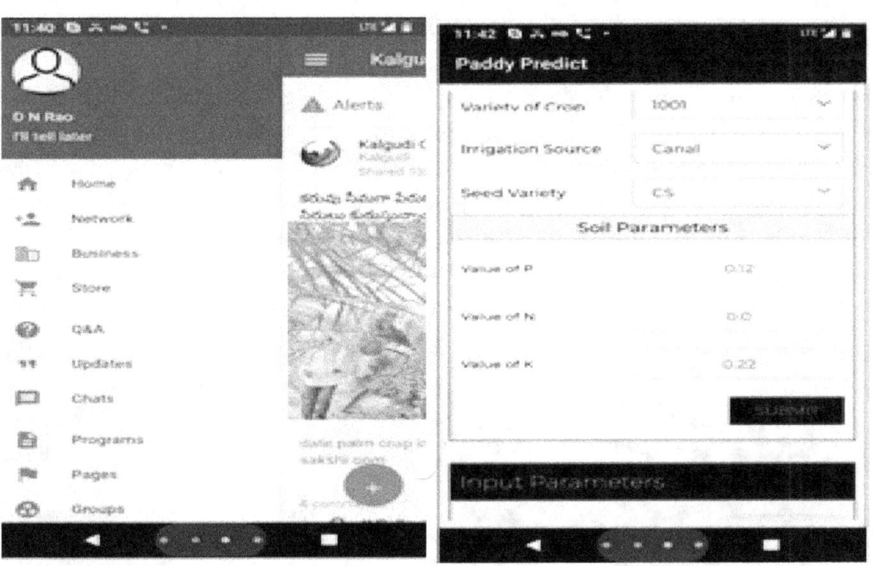

Soil and Crop Health Monitoring using Image Processing

The images were acquired by the Daedalus sensor aboard a NASA aircraft flying over the Maricopa Agricultural Center in Arizona.

Reflectance and Temperature Measurements

The bottom image shows where crops are under serious stress, as is particularly the case in Fields 120 and 119 (indicated by red and yellow pixels). These fields were due to be irrigated the following day

Internet of Things (IoT) Technology

Internet of Things (IoT) technology is expected to play a significant role in sustainable enhancement of agricultural productivity in the real future. Smart Agriculture incorporates IoT based advanced technologies and solutions to improve operational efficiency, maximize yield, and minimize wastage through real-time field data collection, data analysis, and deployment of control mechanism. Diverse IoT based applications such as variable rate technology, precision farming, smart irrigation, and smart greenhouse are expected to group instrumental enhancement of the value of agricultural processes. IoT can also address agriculture-based issues and increase the quality and quantity of agricultural production, making farms more intelligent and more connected. The smart agriculture market is estimated to grow from USD 13.8 billion in 2020 to USD 22.0 billion by 2025, at a CAGR of 9.8%. Key factors accelerating the smart agriculture market growth are the surging use of modern technologies in agriculture farms, growing income levels and demand for protein-rich aqua food, the increasing focus of farmers on livestock monitoring and disease detection, and the increasing emphasis on reducing the management cost by adopting advanced livestock monitoring products. Increasing investment and rising R and D expenditure in agriculture technology across the world and increasing the popularity of land-based recirculating aquaculture systems are expected to offer substantial opportunities for the smart agriculture market.

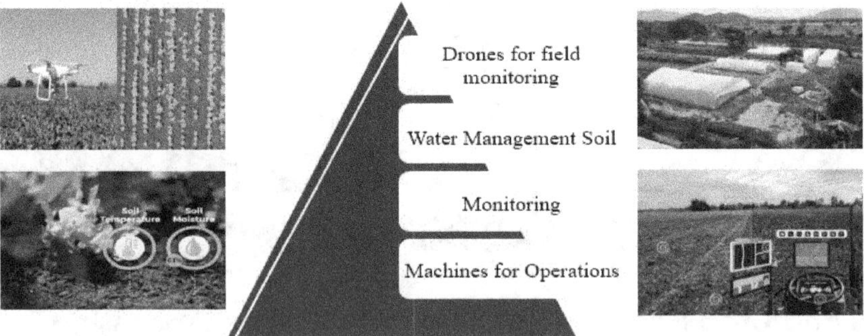

Figure 3.6 IoT – Connected Agriculture

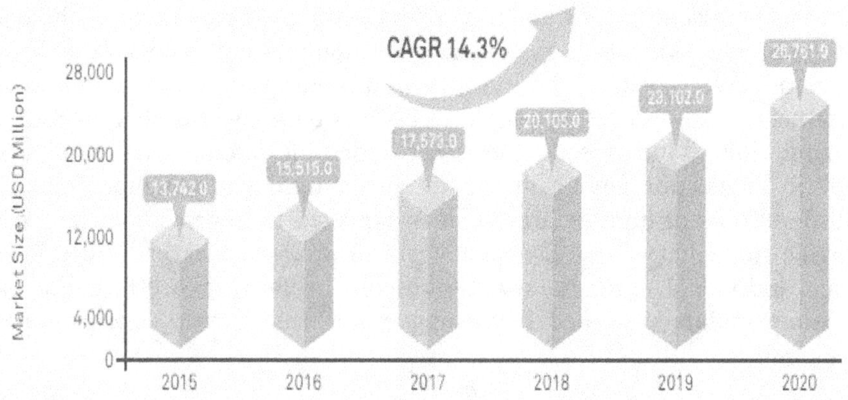

Figure 3.7 Smart agriculture market size forecast (2015–20)

Source: Smart Agriculture Market Assessment, Huawei

AI in Agriculture Industry

AI technology is supporting different sectors to boost productivity and efficiency. AI solutions are assisting in overcoming the traditional challenges in every field. Likewise, AI in agriculture is helping farmers to improve their efficiency and reduce the environment hostile impacts on the environment. The agriculture industry strongly and openly embraced AI into their practice to change the overall outcome. AI is shifting the way our food is produced where the agricultural sector's emissions have decreased by 20%. Adapting AI technology is helping to control and manage any uninvited natural condition. Today, the majority of start-ups in agriculture are adapting AI-enabled approach to increase the efficiency of agricultural production.

AI can improve agricultural productivity; it can identify diseases in plants, it can recognize crop diseases and pest damage, the success was that AI can identify a disease with 98% accuracy, gives growers a weapon against cereal-hungry bugs, Sensors monitor the fruit's progress toward perfect ripeness, adjusting the light to accelerate or slow the pace of maturation. This kind of farming requires considerable processing power.

AI-powered technologies are used in several industries in the world today; finance, transport, energy, healthcare and now agriculture. Farmers can monitor the well-being of their crops or the movement of their animals without the need of going to the farm. Agriculture is undergoing a radical transformation thanks to advancement in AI technology, robotics and AI are reordering what we know about agriculture, making it less labor intensive, and more efficient.

AI-powered solutions will improve quality and ensure a faster go-to-market for crops; Agriculture is both a major industry and foundation of the economy. Many factors such as climate change, population growth and food security concerns have propelled the industry into seeking more innovative approaches to protect and improve crop yield. AI offers data sources such as temperature, precipitation, wind speed, and solar radiation, along with comparisons to historic values for anywhere on the agricultural earth; although AI may not substitute jobs of human farmers, it is expected to improve their processes and provide them with more efficient ways to produce, harvest and sell essential crops. Individual agricultural activities on the farm take effort. For example, planting, maintaining, and harvesting crops need money, energy, labour and resources.

AI in the Agricultural Industry

The most popular applications of AI in agriculture broadly fall into three categories:

- **Agricultural Robots:** Companies are developing and programming autonomous robots to handle essential agricultural tasks such as harvesting crops at a higher volume and faster pace than human labourers.

- **Crop and Soil Monitoring:** Companies are leveraging computer vision and deep-learning algorithms to process data captured by drones and/or software-based technology to monitor crop and soil health.

- **Predictive Analytics:** Machine learning models are being developed to track and predict various environmental impacts on crop yield such as weather changes.

Farmers are increasingly using sensors and soil sampling to gather data and this data is stored on farm management systems that allow for better processing and analysis. The availability of this data and other related data is paving the way to deploying AI in agriculture.

AI Applications in Precision Agriculture

AI is seen as an instrument of transforming rural farmlands into smart connected farms, what we call as Precision Agriculture, is made possible through a combination of smart technologies such as AI, Big Data, Cloud, Internet of Things (IoT), and Machine Learning. Its applications are widely felt - from automatic detection of drought patterns to tracing the ripening patterns of apples or tomatoes, and we now have smart tractors that weed out the diseased and sick plants. Today, even drones are extensively used in agriculture for research analysis, safety, rescue, terrain scanning, spatial analysis, monitoring soil hydration, identifying yield problems, etc.

These smart drones even pin out and precisely spray pesticides on the diseased plants from the vast expanse of farmlands, help add micro and macronutrients, check physical properties such as moisture, chemical properties, pH balance by adding lime, etc. Precision Agriculture along with an AI-powered application facilitates in identifying the match case – tell what disease has crippled the plant and later, match it from the list of its disease imagery database, send corrective measures and so on. So, the possibilities with AI and data analytics seem endless in Precision Agriculture. The collected and analyzed data will be sent as per the farmers' requirements and needs.

The farm-firsts who adopt AI technology into their agricultural practices will undoubtedly gain a significant advantage. Smart technologies can enhance farm yields. The process involves staging multiple data points in large agricultural farms and by collecting precise and pertinent information from multiple data points and various edge devices like drones, sensors, and smart cameras, etc. powered with some intelligent monitoring and analytics systems that finally helps in providing smart information to the farmers.

AI Combines with other Emerging Technologies

AI combined with Machine Learning (ML), Big Data Analytics and IoT-powered smart edge devices like GPS, drones, sensors, radio frequency identification and LED lights are widely used in livestock monitoring, fish farming and smart greenhouses.

What's even interesting is the adoption of the revolutionary Blockchain technology in the agriculture space. Blockchain is now ensuring transparency, faster processing time, zero-time lag whilst upholding superior accountability, all attributes contributing to a thorough optimization of farmers' supply chain produce adding real value.

Technologies that Help Maximize Farm Yield

Farmers can turn their dreams into the reality of assured increased farm output of up to 5-6 times per acre by embracing and adopting future smart technologies into agricultural practices thereby, making the most out of every acre. By any standard, it is a whopping figure. Like never before, today's technology is empowering farmers to take a giant leap forward, thereby dramatically improving their quality of life. From a farmer's point of view, it's a big miracle happening over their small farm place.

We are evolving from the organic food culture to AI-enabled smart precision agriculture and smart connected farm culture where technology dictates and empowers a farmer in a truly rewarding way. Since ages, the farmers world-over, especially in Asian countries, have been waging a long-fought battle with all sorts of external and internal factors such as unpredictable hostile weather, pest problems, water shortage or even drought. Now, the trend can be reversed with AI and Big Data Analytics, which has emerged as a true revelation benefitting the farmer community. It optimizes every inch of the farmland and technically implies that no part of the farmland goes unused or is left wasted as it happened with traditional farming. Today's smart technology ensures every plant is scanned for health and growth tracking, and any persisting pest problems are identified and notified to the farmer - something which was not possible with traditional agriculture methods.

The Future of Agriculture will be Technology-Empowered

The AI technology has become so advanced that nowadays we have tractor machinery that weeds out the infected plants and satellite imagery giving pictures of drought patterns. AI-powered plant app 'Plantix' proved to be a true blessing for the farmers. It is a plant disease and diagnostic tool, which furnishes complete information on plants. It helps in identifying the plant disease and suggests corrective measures to be taken by farmers. Plantix is a mobile crop advisory app for farmers, extension workers and gardeners. Plantix was developed by PEAT GmbH, a Berlin-based AI startup. It can diagnose pest damage, plant disease and nutrient deficiencies affecting crops and can offer corresponding treatment measures.

So, forget the naysayers! It's already happening, the future of agriculture is in the hands of smart technologies and machines, and the early adopters will surely be rewarded. Although the cost part is not yet clear but one thing is for sure, with more and more technology players entering into the market, the growing competition brings efficacy and cost reduction resulting in mass adoption, wherein, it's not just the farmers, but we all will benefit from the agricultural outcomes and food productivity. Farmers can not only get rid of their long-standing agricultural woes but also reap huge profits from the limited farmland they have, by turning them from 'small' to 'smart.'

But, to turn this into reality, the industry thought - leaders, technology companies and agriculture experts should collaborate and work continuously to improve, innovate and harness the potential of AI. Also, the governments across the globe should spread the message on how important Precision Agriculture is to the farmers by educating and encouraging them to embrace emerging technologies like AI, which is proving to be a game-changer having tremendous potential to improve farm yield productivity.

AI in Agriculture Innovation

Agriculture is seeing rapid adoption of AI and Machine Learning (ML) both in terms of agricultural products and in-field farming techniques. AI technology is supporting different sectors to boost productivity and efficiency. AI solutions are assisting overcome the traditional challenges in every field. Likewise.

However, the majority of start-ups in agriculture are adapting AI-enabled approach to increase the efficiency of agricultural production. The global AI in Agriculture market size is expected to reach 1550 million US$ by the end of 2025. Implementing AI-empowered approaches could detect diseases or climate changes sooner and respond smartly. The businesses in agriculture with the help of AI are processing the agricultural data to reduce the adverse outcomes.

The challenges of AI during COVID-19 pandemic

The global health crisis has led to market, supply chain and trade disruptions. Protecting livelihoods and supporting Small and Medium Enterprises (SMEs) and agri-businesses will be critical to maintaining the stability of food systems. The evolving COVID-19 pandemic has made data, analytics and information sharing across food systems more indispensable than ever before.

Medium and Long Term Solutions

AI technologies can be used to address notable challenges in the sector including market fluctuations, irregular irrigation, eroding soil health and sub-optimal pesticide use. Integrating advanced AI technologies in agriculture is required to build the scope and pace necessary for a digitally-enabled transformation of the sector. There is an urgent need for Integration between Agriculture Reform and AI Policy in India.

The opportunities of AI

Globally, enough food is available. However, measures need to be taken to keep supply chains moving, efficient and effective:

- Early warning systems, data and analytics on production, consumption and supply chain is essential to ensure that food and nutrition security goals are met and supply chains are not disrupted

- Most urgently, data and analytics are needed to identify hotspots and bottlenecks in food supply chains to enable supporting action needed

Medium and Long Term Solutions

The use of AI and related technologies have the potential to impact productivity and efficiency at all of the stages of the agricultural value chain. In agriculture, enhanced farmers' income, increased farm productivity and reduction of wastage have been identified as targets for AI-led innovations.

AI technologies can be used to address notable challenges in the sector including market fluctuations, irregular irrigation, eroding soil health and sub-optimal pesticide use. A few priority applications of AI in agriculture in the current scenario are around: Estimate sowing area for required crops, crop yield estimation, a stress prediction model for crops, price forecasting, food grading, precision spraying, AI-powered advisory to farmers.

Current Impact of AI

- To Identify disruptions across the value chains.
- Enable information sharing across the food value chains.
- Protect livelihoods.

Medium and Long Term Solutions

- **Farmers:** Increase profitability to farmers by reducing the input costs, increasing the yield and providing market linkages by deploying innovative IT solutions.

- **Government:** Make an impactful intervention to reduce the stress of the farmers, by taking up a data-driven approach and use all alternative technologies to get real time information on sown acreage, crop stage and estimated yield all of which were collected and reported through field-level interventions.

ICT and Agriculture

Farming and Information Technology are perhaps the most distantly placed knowledge sets in the world, farming being the oldest and most basic source of livelihood and IT being the most advanced and most modern. The importance of harnessing developments in IT for the betterment of farming is now being increasingly realized. The information related to policies and programs of the government, schemes for farmers, institutions through which these schemes are implemented, new innovations in agriculture, Good Agricultural Practices (GAPs), institutions providing new agricultural inputs (high yielding seeds, new fertilizers, etc.) and training in new techniques, is now being disseminated to farmers through the use of IT to ensure inclusiveness and to overcome the digital divide. Soil Management, Water Management, Seed Management, Fertilizer Management, Pest Management, Harvest Management and Post-Harvest Management are important components of e-Agriculture where technology aids farmers with better information and alternatives. It uses a host of technologies like Remote Sensing and Computer Simulation, helps in the assessment of speed and direction of the wind, soil quality assays, crop yield predictions and marketing. e-Agriculture is part of a mission mode project, included in the National e-Governance Plan (NeGP) in an effort to consolidate the various learnings from the past, integrate all the efforts currently underway, and upscale them to cover the entire country.

Emerging Agriculture Technologies

Venture capitalists have invested billions of dollars in agriculture technology start-ups since 2014 and the trend is expected to continue in the future because the demand for innovative farm technology is high. Modern farmers have amply demonstrated their willingness to embrace new inventions and techniques. Some emerging technologies, among many more that can literally change the agricultural landscape in the years ahead, are:

Soil and Water Sensors

Soil and water sensors are durable, unobtrusive and relatively inexpensive. Even family farms are finding it affordable to distribute them throughout their land, and derive numerous benefits. For instance, these sensors can detect moisture and nitrogen levels, and the farm can use this information to determine when to water and fertilize rather than rely on a predetermined schedule. That not only results in more efficient use of resources and therefore lowered costs, but also helps the farm become more environmentally friendly by conserving water, limiting erosion and reducing the levels of discharge of fertilizer into local rivers and lakes.

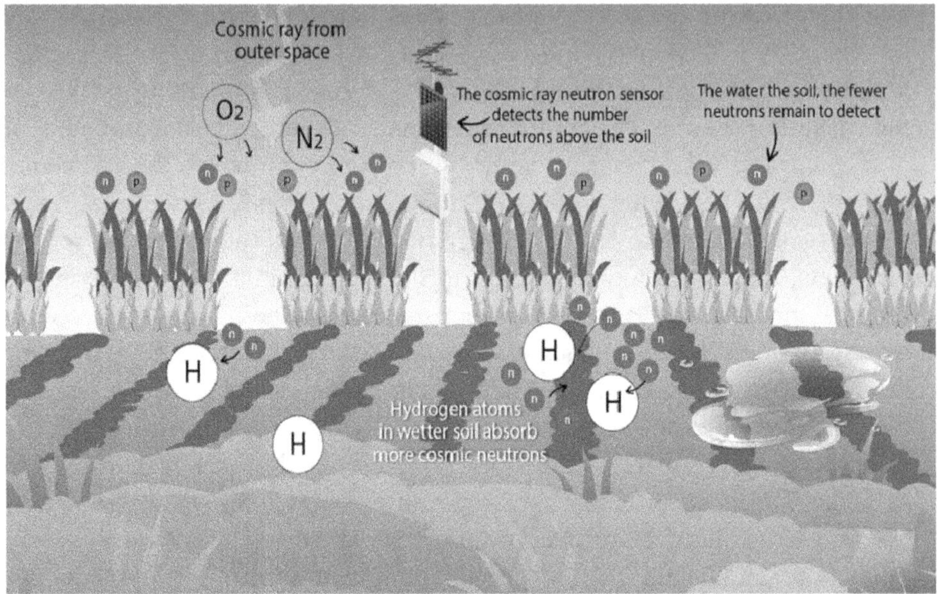

Weather Tracking

Computerized Weather Modelling is becoming increasingly sophisticated. There are online weather services that focus exclusively on agriculture, and farmers can access these services not only on dedicated on board and handheld farm technology, but also via mobile appliances that run on just about any consumer smartphone. This technology can provide advanced notice of frost, hail and other weather conditions to enable precautions to be taken to protect the crops and mitigate losses.

Satellite Imaging

As remote satellite imaging becomes more sophisticated, real-time crop imagery has become possible, providing not just bird's-eye-view snapshots, but images in resolutions of 5-meter-pixels and even greater. Crop imagery lets a farmer examine crops as if he or she were standing there. Even reviewing images on a weekly basis can save a farm a considerable amount of time and money. Additionally, this technology can be integrated with crop, soil and water sensors so that the farmers can receive notifications along with appropriate satellite images.

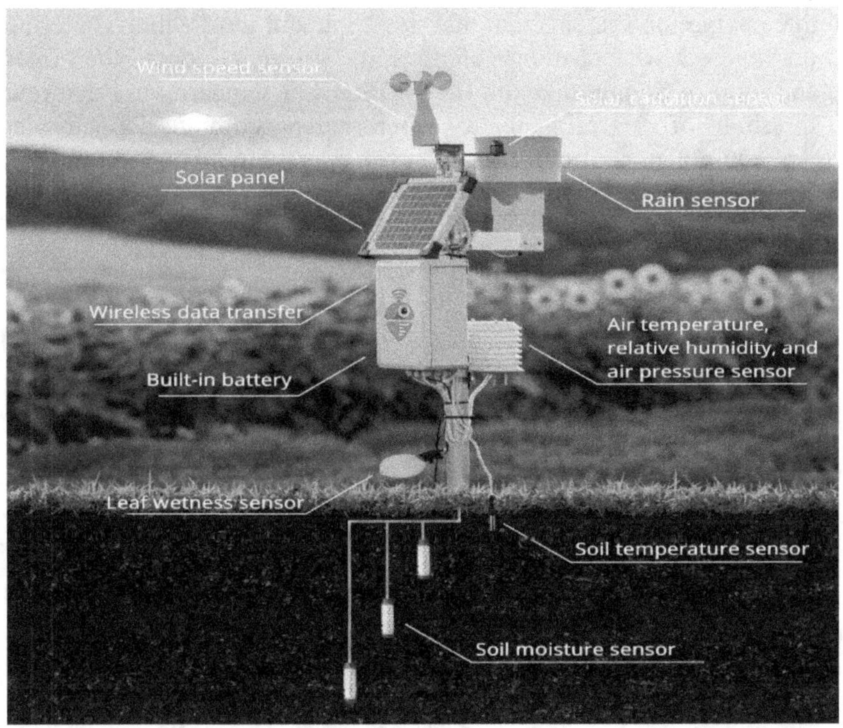

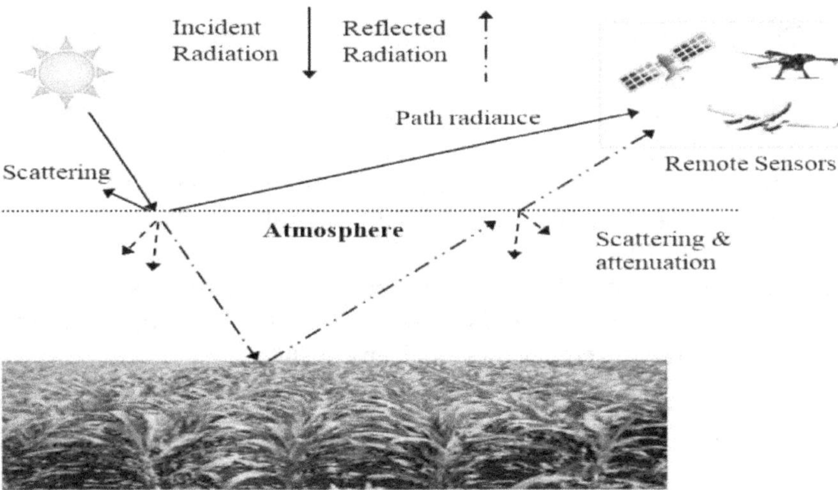

Pervasive Automation

Pervasive automation is now a buzz word in the agriculture technology industry. It can refer to any technology that reduces operator workload. Examples include autonomous vehicles controlled by robotics or remotely through terminals and hyper precision, such as Real-Time Kinematic (RTK) satellite navigation systems that make seeding and fertilization routes as optimal as possible. Most farming equipment already adopts the International Organisation of Standardization (ISO), ISOBUS standard, and that puts one on the threshold of a farming reality where balers, combines, tractors and other farming equipment communicate and even operate in a plug-and-play manner.

Mini Chromosomal Technology

Perhaps one of the most exciting products in agriculture technology comes in a very tiny package. A mini chromosome is a small structure within a cell that includes very little genetic material but can, in layman's terms, hold a lot of information. Using mini chromosomes, agricultural geneticists can add dozens and perhaps even hundreds of traits to a plant. These traits can be quite complex, such as drought tolerance and nitrogen use. However, what is most intriguing about mini chromosomal technology is that a plant's original chromosomes are not altered in any way. That results in faster regulatory approval and wider, faster acceptance from consumers.

Radio Frequency Identification Technology

The soil and water sensors mentioned earlier have laid the foundation for traceability. They provide information that can be associated with farming yields. Theoretically a bag of potatoes can have a barcode that you can scan with your smartphone in order to access information about the soil that yielded them. A future where farms can market themselves and have loyal consumers track their yields for purchase no longer appears far-fetched.

Vertical Farming

Vertical Farming has been a sci-fiction topic from as long ago as the 1950s. Now it is not only scientifically possible but will soon be financially viable. Vertical farming as a component of urban agriculture is the practice of producing food in vertically stacked layers. This offers many advantages. Perhaps the most obvious is the ability to grow within urban environments and thus make fresher foods available at a faster pace and lower costs.

Thus, it appears as though it will not be limited to just urban environments as initially expected. Farmers in all areas can use it to make better use of available land and to grow crops that would not normally be viable in those locations. Technology is transforming nearly every aspect of our modern lives. Farming is no exception. The produce on your table tonight will reach there faster, fresher and more cost-effectively thanks to leading-edge technologies.

Precision Livestock Farming (PLF)

Precision Livestock Farming (PLF) is the use of advanced technologies to optimize the contribution of each animal. Through this "per animal" approach, the farmer aims to deliver better results in livestock farming. Those results can be quantitative, qualitative and/or addressing sustainability.

PLF, or so-called 'Precision Livestock Farming', adds advanced technologies and software to the farming process, with the goal of optimizing the contribution per animal. But to really get the most out of Precision Livestock Farming, a holistic approach is required to overcome less effective isolated solutions. This connects the main areas of focus: nutrition, health, and farming. Therefore, we want to ensure the welfare of the animals and the discharge of the farmers.

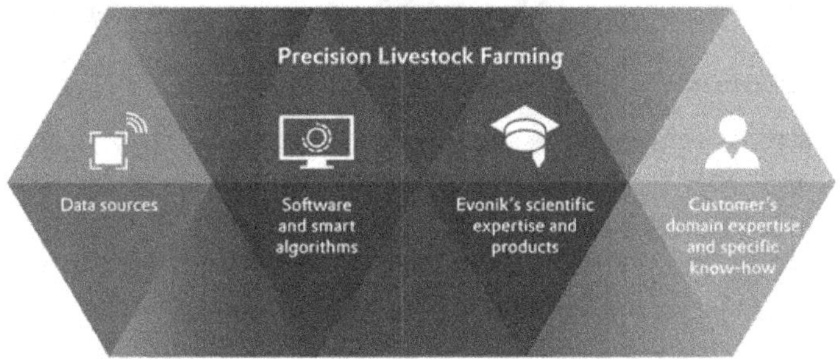

Soil and Crop Health Monitoring using Image Processing, Reflectance and temperature measurements

Continued deforestation and degradation of soil quality are becoming a big challenge for food producing countries. But now a German-based tech startup PEAT has developed a deep learning-based application called Plantix that can identify the potential defects and nutrient deficiencies in the soil including insects and diseases. Plantixis a mobile crop advisory app for farmers, extension workers and gardeners. Plantix was developed by PEAT GmbH, a Berlin-based AI startup. It can diagnose pest damage, plant disease and nutrient deficiencies affecting crops and can offer corresponding treatment measures.

This app is working on image recognition based technology and one can use smartphone to capture the plant's image and detect the defects in the plants. One will also get soil restoration techniques with tips and other solutions on short videos on this app.

Similarly, Trace Genomics is another machine learning-based company provides soil analysis services to farmers. Such apps help farmers to monitor the soil and crop's health conditions and produce a healthy crop with a higher level of productivity.

SkySquirrel Technologies developed by another similar company Vine View brought drone-based aerial imaging solutions for monitoring crops health. A drone is used to make a round of capturing the data from the vineyard field and then all the data is transferred via a USB drive from the drone to a computer and analyzed by the experts.

The company uses the algorithms to analyze the captured images and provides a detailed report containing the current health of the vincyard, generally the condition of grapevine leaves as these plants are highly prone to grapevine diseases like molds and bacteria, helping farmers to timely control using the pest control and other methods.

Use of AI in Harvesting of Crops

AI is expanding its footprints at the ground level, making a significant impact in the world's most vital sector -Agriculture. After healthcare, automotive, manufacturing and finance sectors now AI in agriculture is providing cutting-edge technology for harvesting with better productivity and crop yield.

The Agriculture sector is the foundation of the world's economy and with the increasing population, the world will need to produce 50% more food by 2050. AI-enabled technologies can help farmers get more from the land while using resources more sustainably. Here, we'll learn how AI can be used in agriculture and its applications in farming.

Autonomous Tractors

With the heavy investment in developing autonomous vehicles for various needs, the agriculture sector will also be getting benefits with self-driving or say driverless tractors.

With more quality AI and machine learning training data for agriculture, the farm sector is going to be revolutionized by the large scale use of autonomous tractors for performing multiple tasks.

These self-driving or driverless tractors are programmed to independently detect their ploughing position in the fields or decide the speed and avoid obstacles like irrigation objects, humans and animals while performing various tasks.

Agricultural Robotics

AI companies are developing robots that can easily perform multiple tasks in the farming field. Such robotics machines are trained to control weeds and harvest the crops at a much faster pace with higher volume compare to humans.

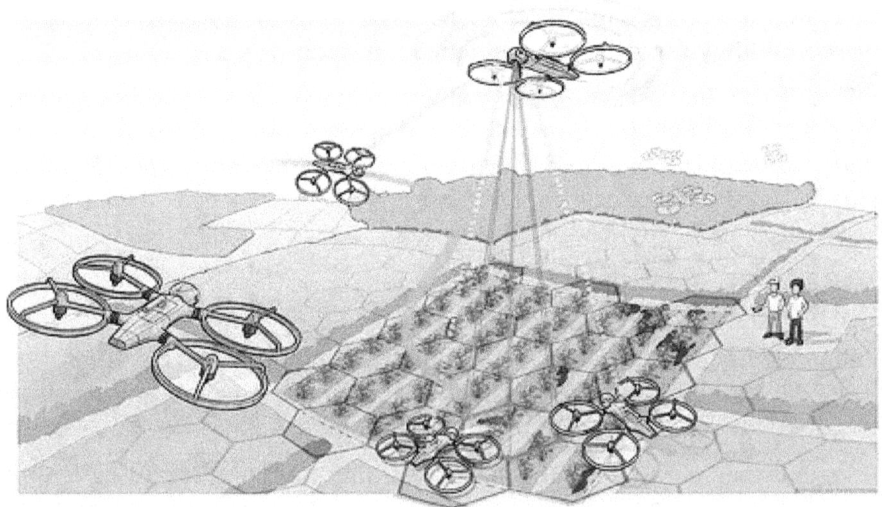

These robots are well-trained to assist in checking the quality of crops and detect unwanted plants or weeds with picking and packing of crops at the same time capable of fighting with other challenges faced by the agricultural labour force.

Companies like Blue River Technology and Harvest CROO Robotics are making robotics machines that can control unwanted crops or weeds and help farmers in picking or packing of crops with higher volumes.

Controlling Pest Infestations

Pests are one of the worst enemies of the farmers damaging the crops globally before they are harvested and stored for human consumption. Popular insects like locusts, grasshoppers, and other insects are eating the profits of farmers and gobbling the grains meant for humans. But now AI in farming gives growers a weapon against such bugs. Plantix Pest Tracker since 2018 is live monitoring an invasive pest Fall Army Worm (FAW).

AI and data companies are helping farmers to get alerts on their Smartphones about the grasshoppers likely to descend towards a particular farm or matured crop field.

AI companies using the new satellite images against pictures of the same using historical data and AI algorithm detects that the insects had landed at another location and farmers use such information after confirmation and timely remove the costly pests from their fields.

Precision Farming with Predictive Analytics

AI applications in agriculture expanded into doing accurate and controlled farming through providing proper guidance to farmers about optimum planting, water management, crop rotation, timely harvesting, nutrient management and pest attacks.

While using the machine learning algorithms in connection with images captured by satellites and drones, AI-enabled technologies predict weather conditions, analyze crop sustainability and evaluate farms for the presence of diseases or pests and poor plant nutrition on farms with data like temperature, precipitation, wind speed, and solar radiation.

AI in agriculture is not only helping farmers to automate their farming but also shifting to precise cultivation for higher crop yield and better quality while using fewer resources.

Companies involved in improving the machine learning or AI-based products or services like training data for agriculture, drone and automated machine making will get technological advancement in the future will provide more useful applications to this sector helping the world deal with food production issues for the growing population.

Promoting Continuous R and D

Every day, millions of rural people who depend on agriculture confront technical, economic, social, cultural, and traditional obstacles to improving their livelihoods. To cope with these obstacles, the rural poor draw on the legacy of inherited Indigenous Technical Knowledge (ITK) and innovate through local experimentation and adaptation. ITK alone, however, is not enough to deal with the complex problems facing the agricultural sector. Emerging issues such as volatile markets, climate change, and demands for biofuels require complementary knowledge from formal R and D, and support from policies and institutions. Formal and informal knowledge must be created, accumulated, shared, used and the innovations to be acquired which can be shared either in the form of knowledge or information to farmers. And for new ideas, practices, or products that are successfully introduced into economic or social processes - can involve technologies, organizations, institutions, or policies. Advancing agricultural development requires knowledge and innovation in several key areas:

Technology: While many good technologies are "on the shelf," emerging issues such as climate change require new research to develop drought-resistant, flood-resistant, and short-duration crop varieties.

Institutions: More socioeconomic research is needed to understand institutional constraints in the effort to improve livelihoods. Institutions are the arrangements, systems, mechanisms and rules that constitute the environment within which innovations occur and include laws, regulations, traditions, customs, beliefs, norms, and nuances of society.

Policies: Appropriate, relevant, and timely public interventions are needed to promote and facilitate the creation, sharing, and use of knowledge for innovations.

Organizations: Public and private groups and companies must innovate to become more effective and efficient in the services they provide.

To foster innovations in agriculture, policymakers must scale up investments in agricultural science and technology, research and extension, agricultural education and training, and farmer organizations and other local institutions—and do so in ways that will spread advances in knowledge and innovation as widely as possible.

Key Policy Options for Promoting Knowledge and Innovation for Agricultural Development: Governments of developing-countries usually face policy choices, and, given their limited resources, they need to make decisions carefully. Pragmatic policies and government actions can encourage actors in the food and agriculture value chain to create, accumulate, share, and use knowledge.

Farmer Centred R and D: Policy and institutional innovations can be used to motivate the private sector to undertake or finance agricultural research and to encourage commodity associations to allocate funds to research institutes or universities for commodity research. Policymakers can also provide incentives for agro-processing firms to establish laboratories to carry out or finance food research. Other policy innovations can include competitive grant schemes to direct research into areas of immediate need or prizes for outstanding research. Organizational innovations in research organizations can include a participatory research approach that brings in all of the actors in the food and agriculture value chain. Working collaboratively, national agricultural research institutes, international research centers, farmers, and extension services have already produced numerous research results that have led to increased knowledge and innovation in agriculture. These results show the importance of focusing not just on technical, but also institutional, organizational, and policy innovations in getting research funded, organized, and implemented efficiently and getting the information shared among users to be processed for innovations.

Extension to Take the
Technical Improvement to Farmers' Doors

Agricultural Extension: Agricultural extension is an important player that can bring together research, farmers, and other players in the innovation system. The extension is defined as the set of services that supports people engaged in agricultural production to help them solve problems and obtain knowledge, information, skills, and technologies to improve their livelihoods and well-being. Extension approaches have evolved from departments of government, national extension systems, the Training-and-Visit (T and V) system to privatized (and otherwise reformed) systems. What matters is not so much the approach or system, but its ability to offer a "best-fit" solution to local needs and conditions. Technologies, information, and skills are of no use if they do not take users into account or do not reach the farmer's needs. Many farmers complain about the ineffectiveness of extension services, which are viewed as supply-driven, highly centralized, non-participatory systems that exclude the poor. For example, worldwide, women farmers receive only 5 percent of extension services, whereas research has shown that farm productivity increases by 22 percent when women receive the same advisory services as men. The public extension must bring about technical, institutional, and organizational reforms to make it more cost-effective, demand-driven, and participatory. Advances in information and communication technologies offer opportunities for technical changes from which both extension staff and their clientele can benefit. Mobile phones and Internet kiosks provide quick and affordable channels for relaying agricultural advice. In Kenya, a new system that reads out text via mobile phone is helping banana producers. This Banana Information

Line, a boon to illiterate farmers, is available in Kiswahili and English and helps users troubleshoot banana cultivation problems. In Sierra Leone, mobile phones are supplied by a project on ginger link agronomists and extension workers. Providing mobile phones to agronomists, extension workers, and farmers can be a cost-effective means of sharing information among these three groups. Within extension, institutional and organizational changes are also required, so that the service goes beyond technology transfer to facilitation and beyond training to learning. It must include processes such as assisting with the formation of farmer groups, dealing with marketing issues, and partnering with a broad range of service providers and agencies.

Education and capacity development: Capacity development is a key policy priority for stimulating knowledge and innovation. Innovation system actors and organizations require strengthening at many levels in order to work more effectively. Farmers and farmer organizations require a strengthening-for instance; establishing successful demand-driven extension services requires strengthening users' capacity to demand the types of services they need. Organizations (public sector, private sector, or civil society) that provide extension services demanded by users must also be trained to respond to users' needs. Researchers need to learn how to work with farmers and communicate with extension workers. And policymakers need to better understand the innovation system and its different components. For decision makers, the key question is where to spend scarce resources on capacity strengthening: at the government level, at the level of research, education, and extension organizations, or among farmers and their organizations? Priorities depend on local conditions, such as socioeconomic level, governance structures, political system, and availability of infrastructure. All stakeholders need to participate in setting priorities.

Organizational and Institutional Innovations: Organizations and institutions that guide the performance, outcomes, and impact of the agricultural sector must be more innovative to be more efficient and effective. National research organizations, extension organizations, community and farmer-based organizations, and rural service providers within the food and agricultural value chain must be strengthened to enable them to innovate and function efficiently. To maximise their impact, these organizations must be able to set long- and short-term strategic objectives, set priorities, and establish an organizational performance assessment system to guide the innovation processes they are involved in.

Sustainable Farm Livelihoods

Farmers around the world are adopting better technology at an accelerated rate; individuals and organisations are to help farmers to increase crop yields, protect soil conditions and maximize profitability so they can thrive. Farmers are using tools that increase planting efficiency and allow more precise application of crop inputs, such as fertilizer, thereby boosting yields while reducing waste and environmental impact. Individuals and Corporate Social Responsibility (CSR) of organisations are helping farmers at all levels of productivity to be equipped with the knowledge to improve yields sustainably, provide reliable markets for their crops and help them manage risk. It is essential for these farmers to contribute more fully to helping achieve a more sustainable, food-secure future.

The following are broad demands laid down by individuals and organisations working towards sustainable farm livelihoods and farmers rights:

- Periodic safety review of registered pesticides to be undertaken every five years

- Adequate representation and power to state governments. States like Telangana, Maharashtra, Punjab and Kerala wanted to ban a particular herbicide like glyphosate in their respective states but were not allowed to do so for a long time.

- Mandatory recoding of data by state government related to farmer deaths due to poisonous inhalation.

- Compensation funds to be set up for affected farmers need to be made more sound in terms of legal groundwork.

- Collection of assets from the pesticide industry for compensation to be provided to farmers.

- Ban on all kinds of advertisements and promotions of pesticides.

- Empower the pesticides management board being set up as a regulatory body instead of an advisory panel.

- Review committee for pesticides to be constituted independent of the Registration Committee.

Food Supply Chain

The food supply chain is linked with different actors from 'farm to fork' to achieve a more effective and consumer-oriented flow of products. Supply chains include growers, pickers, packers, processors, storage and transport facilitators, marketers, exporters, importers, distributors, wholesalers, and retailers. Supply chain development can thus benefit a broad spectrum of society, rural and urban areas. Managing supply chains requires an integral approach in which chain partners jointly plan and control the flow of goods, information, technology and capital from 'farm to fork', import from the suppliers of raw materials to the final consumers and vice versa. Conversely, digital transformation in the supply chain is disrupting agriculture as follows:

Disrupting Agriculture

- Assisting research and development for the improvement of seed quality, crop yield, effective fertilizers and pesticides.

- Modelling and simulation, analysis of genetics, genomics and phenomics.

- Improving farm equipment's through design inputs.

- Promoting protected and vertical agriculture.

- Soilless and waterless agriculture (hydroponics and aquaponics).

- Blockchain based agriculture collateral management and credit services.

Digital Soil Information for Farmers/Landlords on Mobile App

- Updated Fertility Info of soil-based on image analytics (no need for soil testing).

- Crop suggestion, yield projection based on analytics, machine learning principles.

- Guidance on Fertilizers/Pesticides and crop package of practices customized to farmers land.

- Precision fertigation and pesticide use using drones and IoT devices.

Agriculture Digital Services

- Single window project progress dashboard for all agri-related initiatives of Government.

- Tapping the info from each initiative specific site or database.

- Value add services on e-NAM, e-KISAN and other government initiatives.

- Reorienting agriculture extension to provide business advice to farmers (Krishi Udyog Sahayak).

- Using AR/VR for farmers training.

- Training for farm as enterprise mindset creation.

Center of Excellence for Skill Enhancement, Self-Employment, Farm Enterprise Incubation

- Training on modelling, simulation of genetics, genomics and phenomics

- Software, hardware infrastructure for training and innovation/incubation of ideas.

- Management of Innovations in warehousing and Supply chain

- Online marketplace support for custom hiring farm equipment and agriculture inputs.

CHAPTER - 4

THE ROAD AHEAD

With the rising population, food insecurity and malnutrition have become serious threats. The world's population is expected to reach 8.5 billion by 2030 and 9.7 billion by 2050; therefore, it is necessary to take measures to tackle these growing threats. Agriculture is the foundation for ensuring food security, and therefore, has been the focus of policy formation.

In India too, over the past decades, various reforms have been undertaken to harness the potential of the Indian agriculture sector fully. Agriculture is the key driver of the Indian economy and continues to be the source of income for the majority of its population. Therefore, these reforms have been in the form of favourable policies, infrastructure development, extension and knowledge support, financial and input support. However, certain challenges persist in the sector, such as low productivity, sub-optimum price realisation by farmers, lack of market connectivity, post-harvest losses, etc. In a recent initiative to empower farmers, the Government of India presented the agenda of "Doubling Farmers' Income" by 2022. The initiative focuses on farmers at its core and requires strengthening measures across sub-sectors.

In transforming India's agriculture, there will be basic challenges, which are structural in nature. The planned change has to be time-bound, resource use efficient and cost effective. In order to drive the change for desired results, the speed & quality of implementation of the action plan is important. However efficient such execution, the outcome will be limited, if the system suffers from certain inherent constraints. These need to be identified and addressed to realise the full potential of the strategy for change. Governments at both central and state levels, as also resource credit institutions, make available support-system in terms of inputs, technology, management practices and knowledge. In the larger interest of agriculture, there is a need to adopt a liberal approach in defining a farmer.

This chapter examines the need of structural reforms, agriculture extension reforms, good agriculture governance, governance reforms and strategies, raising farmers income, innovative policy interventions, strategy for faster and sustainable agricultural growth leading to the doubling of farmers' incomes, way forward, policy interventions, research and development, extension and good governance, farm-to-fork approach for doubling farmer income.

4.1 Structural Reforms

Factors such as rising income levels and wealth, the growing middle class, nuclear families and working couples and increased urbanisation are expected to result in India being a fast-emerging hub for processed foods. India's exports of processed food products have increased steadily from US$ 1,352 million during 2006 to US$ 3,981 million in 2016.

The food industry, which was valued at US$ 39.71 billion as of 2013-14, was expected to grow at a CAGR of 11% to reach US$ 65.4 billion by 2018. Indian food service industry is expected to reach US$ 78 billion by 2018. India's organic food market is expected to increase three times by 2020, as compared to 2017. In the year 2012-13, there were 1.6 million people engaged in the registered food processing sector. This number is expected to increase to 9 million by the year 2024.However, for improving the efficiency of the entire value chain, there is a need for substantial investment in infrastructure development. Some of the investment needs and opportunities for investors are as follows:

Production

- Production of high-yielding seeds.
- Production of high-quality planting material, including the use of tissue culture methods of micro-propagation.
- Nurseries including hardening nurseries (practices during the nursery cycle that prepare plants for the stresses of handling, shipping, out planting, and field establishment).
- Organic farming.

Production of microbial cultures and vermicomposting, and

- Floriculture.
- Shift from mono cropping to diversified crops, crop schedules, crop planning, commodity boards to have control on seed quality, acreage, package of practices, value addition, export competitiveness, better realizations, etc., are essential.
- Seed to plate approach through structured contract farming approach will be helpful to de-risk the farmers.
- Small farmer collectives with pooled inputs sharing, producing organic food with traceability will boost or boast better realizations from upcountry markets and exports. Certification support and simplified systems are essential.
- Post-harvest and production systems based on current global trends, change in food habits will certainly improve the farmers' portfolio. For example,

take the case of sugarcane farmers, sugar prices crashed globally for the last four years and farmer dues from mills are mounting to thousands of crores. If one can understand sweetener consumption patterns globally there is a considerable shift from white crystal sugar to crystal or granulated jaggery. Sugar is selling at ₹ 35/Kilo in retail, whereas granulated jaggery is selling above ₹ 160/kg. Price delta is 4 times over sugar. These types of examples are plenty. India being country with a predominance of small and marginal farmers' country, can easily grab these opportunities and boast our economy.

Finance

- Restructuring loans for the mark to market (MTM) and fresh capital required to address new emerging needs particularly marketing and agroindustry through institutional framework at mandal/block level. To be able to present the consolidated problem and solution to bank and the lead banks.

- Warehouse receipt finance shall be made easily available to farmers, so that they will be able to stock and sell the produce during the lean season.

- The scale of finance for value addition, export targeted ESG compliance production, FDI into such facilities with government guarantees shall be encouraged. We have huge potential in fisheries, meat and other farm-based products.

Processing

- Fruit and vegetable processing, including dehydration, canning, aseptic packaging, processing of underutilized fruits, and processing for other products like grape raisins, osmo air-dried fruits, fruit toffees, bleached dry ginger and spices' powders.

- Processing of maize for starch and feed through improved mini/small mills and dry milling plants.

- Processing of millets for various purposes, including malt from finger millets and RTE (ready-to-eat) products.

- Processing of sugarcane for various jaggery products like spiced jaggery, powdered jaggery, and jaggery cubes.

- Processing of herbal and medicinal plants.

- Processing of dairy products.

- Processing for poultry products, including poultry dressing, and

- Processing of livestock products and livestock wastes.

Infrastructure

- To the extent possible rural godowns, individually owned storage systems (for example in coastal AP in many villages the houses are built in such a way that the ground floor is grain storage and the first floor is the house). Rural banks shall be mandated with financing these facilities and extend warehouse receipt finance to these systems.

- Custom hire centres for farm machinery, innovative weeding and other tools for small farmers, solar drying facilities, village common infra facilities shall be encouraged.

- Facilities such as cold chain, reefer trucks, sea and airport handling facilities, supply chain logistics need considerable investment.

Marketing & Sales

- Diversify the crop mix for biodiversity and market de-risking.

- Target 30% D2C sales local and 70% Bulk sales aligning with private firms (POPI) in the biz of aggregating and supporting FPOs.

- For example, in our village buffalo milk is procured by society @ ₹ 700/kg Fat. With an average 7% fat & 8.5% SNF, the average price realization is ₹ 42/Kg. Whereas, same society is selling milk locally @ ₹ 68/litre. About 60% of the collection is sold locally. After considering ₹ 8/litre as overheads, bonus, etc. net realization/litre is ₹ 18/- (36%). This is the strength of our local economy. If farmers can organize into groups and take-up activities like this, rural economic growth will surpass all sectors.

- Innovative marketing models which are globally successful like sell before you show, transaction models, annual contracting models with global food giants like whole foods, etc., have huge potential.

- Reorienting existing market yards as aggregation points, logistics, insurance, payment, QC facility centres with linkages to farmer groups, integrating with e-commerce will bring transparency in price realization to farmers.

Institutions

- India started very well with PACS, Co-op Federations etc., and created rural infra, market yards, etc., with farmer-friendly democratic approach. Over the year's politics entered cooperatives and the structure took a huge beating.

- Wherever leadership was strong and uncompromising these people's institutions became role models. The best examples are AMUL in Gujarat and Mulkanoor in Telangana. AMUL with over ₹ 70,000 Cr turnover with 3 million farmer base cooperative brought laurels to the nation globally. The success is attributed to two towering personalities Dr. Kurien and A. Viswanatha Reddy.

- After seeing PACS, MACs, Producer companies, now we are promoting farmers producer organizations. Whatever may be the name, the purpose is to protect farmers' interests and to find end-to-end solutions to the farming system.

- At present the institutions like NABARD who are mandated to promote people's institutions, are focusing more on the huge number of FPOs, rather than quality. Consultancy organizations, NGOs who mushroomed during World Bank/SERP days have a new lease of life again with GOI promoted FPOs.

- Many FPOs/FPCs had natural death due to incapable managements, lack of working capital, failure in reaching markets and not in a position to pay statutory filing charges other than huge audit fees. The importance of leadership is ignored in many cases and unfortunately, we are not learning from earlier mistakes.

Micro-irrigation is considered to be one of the most efficient solutions to overcome the water management challenges faced by the agriculture sector. With the government's focus on providing rural electrification, the country is expected to witness a growth in the level of mechanisation in agriculture production. Mechanised agriculture will reduce the operational cost and time by improving post-production agricultural activities and promoting conservation of water. Government initiatives in several areas like promotion of farmer producer companies, organic farming, and better soil management (through soil health cards) and better farmer-market linkages (through e-NAM) are expected to further propel the sector.

Mega Food Park Scheme will create a centralised infrastructure to take care of processing activities which require cutting edge technologies and testing facilities, besides the basic infrastructure for water supply, power, environmental protection systems, communication, etc. The Indian food processing industry will serve as a vital link between the agriculture and the manufacturing sectors of the economy which would further enhance the production, consumption and export in the food market. The food processing industry is expected to grow at a CAGR of 13% during 2015-2020.

Agriculture Extension Reforms

The agricultural extension schemes focus on operationalizing agricultural reforms across the country though new institutional arrangement with restructured autonomous bodies at District/ Block level, which are flexible, bottom-up, farmer-driven and promote public-private partnership. The following participatory bodies intend to set up at various levels:

1. State Level: Inter-Departmental Working Group (IDWG)
   - SAMETI Executive Committee.
   - State Farmers Advisory Committee (SFAC).
2. District Level: ATMA Governing Board
   - ATMA Management Committee.
   - District Farmers Advisory Committee (DFAC).
3. Block Level: Block Technology Team (BTT)

Block Farmers Advisory Committee (BFAC)

In order to achieve the target of doubling farmers income enhance the use of Farmer Interest Groups (FIGs) to mobilize men, women and young people around a common interest, such as the production of flowers, fruits, vegetables, milk, fish and other high-value products, has energised both the farming community and the extension staff. Developing a strong farmers' organisation is a positive and necessary step in providing cost-effective extension services. It will increase the income of farmers and employment of small scale and marginalised farm households. Agriculture extension services need to be made more effective and private extension services must supplement and complement the farmers.

Extension services do not exist for animal husbandry and should be developed at the earliest. Agriculture extension services have to be revitalized by making them more relevant, useful and timely in order to improve agricultural productivity by taking the form of a one-stop-shop that offers both hardware and software solutions to raise the incomes of farmers, especially small and marginal farmers. New technologies can be made effective by leveraging information technology and mobile applications. In view of the larger dissemination of visual media and smartphones in rural areas, these mediums can be effectively utilized for the dissemination of technology and package of practices (Ex. Kisan TV by providing farmers a direct interface with agricultural experts).

Good Agricultural Governance

For sustainable agricultural development, the existence of a predictable and transparent framework of rules and institutions is a sine qua non to implement the conceptualized projects. Good governance, therefore, emphasizes establishing effective mechanisms, processes and institutions through which farmers articulate their interests, exercise their legal rights, meet their obligations and mediate their differences.

Good governance can broadly be assessed in the following terms.

- Voice and accountability: to eliminate the abuse of public office for private gains and diversion of resources.

- The rule of law: access to justice and enforcement of rights to land, water, food, etc.

- Regulatory quality: quality of agricultural development policies and regulations.

- Government effectiveness: efficiency and equity in the provision of agricultural services and infrastructure.

4.2 Governance Reforms and Strategies

1. Governance reforms require changes in institutional structures and legal system, viz.: (i) decentralization of the provision of agricultural services (research, education, extension, input supply, processing, marketing, etc.), functions/responsibilities (staff) and funds (resources) to lower levels of government (district, block and villages) and (ii) institutional reforms of ministries and departments of agriculture (creation of autonomous agencies, seeking the involvement of private sector, development institutions and NGOs and transfer of management functions to local groups). In the current context, the emphasis has to be to significantly strengthen Panchayati Raj Institutions (at district, block and village levels) financially and equip with professional managerial staff to be capable of planning and implementing agricultural development projects.

2. Governance reforms to improve existing legal and institutional framework by (i) Promoting effective participation of farmers (more importantly groups of small, marginal, women and tenant farmers, share croppers and oral lessees) and their organizations, e.g., Consortium of Indian Farmers Association in agricultural policy processes; (ii) Refining management information system in the department of agriculture at local levels; (iii) Initiating community-driven agricultural development projects and (iv) Supporting Self-Help-Groups of small producer-organizations. In the present context, efforts have to be to create awareness among farmers and seek their full participation, train them and build their capacity.

3. Strategies that can lead to governance outcomes are acknowledged as "demand-side" and "supply-side" strategies (i) the "demand-side" strategy aims at strengthening the ability of the farmers and their organizations to demand better public services on time and hold public officials accountable, including elected representatives, legislators and administrators. This strategy refers to the "voice and accountability" dimension of good

governance; (ii) the "supply-side" strategy aims at strengthening the capacity of the administrators and other service providers to supply services effectively and efficiently and to be more responsive to farmer's priorities and needs. This strategy refers to the "Government effectiveness" dimension of good governance.

Decentralization

The decentralization process, as a part of good governance for agriculture, holds considerable promise to provide better services by "bringing Government closer to farmers". Important features of decentralization include 1. Transfer of professional staff, administrative functions and financial resources to lower levels of Government at district, Block and villages; 2. Political decentralization accompanied by fiscal and administrative decentralization has considerable potential making service providers accountable to farmers and locally elected Governments; 3.Active participation and committed consultation with farmers and their interest groups enables Government to develop workable and sound agricultural policies and regulations; 4. Promoting local level farmers' organizations (e.g., Consortium of Indian Farmers' Association) at the district level and strengthening farmer's rights to information can make decentralization effective: 5. Increasing levels of literacy among farmers and women and training them in farm planning and budgeting contributes substantially to improve the effectiveness of this strategy; 6. Democratic decentralization has to continuously improve the engagement of locally elected council members in policy making at lower levels. For this, there is a need to significantly improve the understanding, skill and ability of farmers to visualize demands and involve themselves in policymaking, directly, through interest groups, and through elected representatives; 7. On lines of the national and state budgets, the budget at local levels is crucially important to ensure that policies befitting local needs are actually translated into practice. The budget process can include five-year perspective budget, annual budget, sector-wise investment plans accompanied by quarterly reviews; 8. Right-to-information facilitates farmers to access information and farmers report cards, based on surveys soliciting their reaction and satisfaction with the quality of public service provided, help bring out improvements; 9. It is necessary that the decentralization process should be administratively capable of preventing the local elite from capturing and using public resources to their advantage.

Farmers' Voice

India has over a period of time evolved and put in place a decentralized governance system for agricultural development in specific areas that, inter alia, include agricultural research, education and training, extension services, input supply system, processing, storage and warehousing, agricultural marketing. This has yielded encouraging results in increasing farm productivity, output and reducing rural poverty. Now is the opportune time to build on these successes and draw a road map of good governance, inter alia, concentrating further on decentralization, thereby strengthening organizationally, financially and functionally lower level Governments particularly at Block and village level and demonstrating a commitment to partner and consult farmers and farmers' organizations to know from them what they really want in respect of research, new agricultural development strategy, accessing financial resources, among others.

Innovative Farmers

It has been well documented that there are several innovative farmers. They have keen observation power and better insight about agriculture-related problems. They use traditional knowledge judiciously to fight area-specific problems in agriculture. Their creative potentials have to be acknowledged. Their active participation and consultation can help develop appropriate location-specific technologies. Experience suggests that formal researches often do not address the problems of high-risk areas and, therefore, farmers themselves develop on their own crop varieties and cultivation practices for their survival. Recognizing that *"Necessity is the mother of invention"* farmers have developed manually operated water lifting devices and pumps to lift water from small streams to higher elevations. Farmers in high risk areas are more innovative than other farmers to find a solution for their survival.

Among innovations, noteworthy are plant varieties, cultivation practices, processing, agricultural tools, non-chemical pest management, harvesting, storing, preservations, mechanical technologies, soil and water management, etc. The participatory approach is the appropriate tool for understanding problems and technological gaps in high risk areas. Endeavours should be to develop technology by encouraging innovative farmers. It is a participatory approach and necessitates the involvement of all stakeholders, viz. research institutes, financial institutions, innovating farmers, NGOs, women SHGs, to promote local innovations in a scientific and sustainable way. Active participation of innovative farmers and their organizations in each of the country's agro-ecological regions in decision-making policy can significantly help farmer-friendly designing agricultural researches in areas, *inter alia*, viz. 1. Development of high-yielding varieties of crops; 2. Agronomic practices involving the use of seeds, fertilizers, pesticides, water in different combinations to evolve yield-maximizing and cost-minimizing technologies; 3. Use of farm tools, equipment and machinery; 4. Post-harvest management; 5. Agricultural advisory services for efficient management of land, water and energy resources, agricultural marketing, etc.

There have been over 100 agricultural development schemes framed by the Central Government and are being implemented by State Governments in which farmers in a large number of districts out of 630 need significant changes to suit their areas depending on soil, climate, socio-economic conditions and limited resources, skill and capacity while implementing. Schemes they need should be area-specific and farmer-friendly. Every year Union Budget adds a good number of schemes supported by subsidies and institutional credit which farmers find extremely difficult to implement since they cannot easily and timely access subsidies and credit. They need their participation and consultation to design hassle-free and speedy procedures, methods and systems to access subsidies and credit for implanting schemes.

Farmers need to be given an opportunity to voice their views on important legislation being drafted or enacted by Parliament, viz. different Acts, regulations and innovative policies.

An enactment needs to be put in place in which:

- Farmers can be fully aware of all recommendations including their rights and obligations contained in the Report on" National Commission for Farmers".

- Farmers can insist on their participation and consultation before allocation of funds for agriculture in annual budgets as they feel allocations at the national level are arbitrary and do not reflect the realistic needs of farmers of different agro-ecological regions, States and districts.

- Tenant and women farmers need immediate legislation establishing their legal rights and relationship with the land they cultivate to help them invest, access credit, insurance, etc.

Every rupee spent on agricultural research yields a return varying from ₹ 1.39 to ₹ 3.66 on different food crops. Total Government spending on agricultural research has stagnated over the past decade at meagre 0.6% of the agriculture sector's GDP and is considerably lower at 0.14% on agricultural extension services to transfer technology to farmers. Farmers since long have been insisting to progressively raise to a minimum at 2% and 1% respectively by 2017. Besides, agricultural research institutes and centers need to be widely dispersed in all agro-ecological regions and extension service network in all villages of the country. According to reports total R&D expenditure in India as a percentage of GDP has been stagnant at 0.6 to 0.7 per cent in the last two decades — much lower than the US (2.8 per cent), China (2.1 per cent), South Korea (4.3 per cent) and Israel (4.2 per cent). To keep the numbers in perspective, one must keep in mind that GDPs of the US and China are around seven and four times bigger than that of India. China's spending on agri research is 3-4 times that of India.

As agriculture has been a State subject farmers have serious views that States in general and districts/blocks in particular must be given adequate financial resources on time, hassle-free and full authority and flexibility to frame research policy, formulate development strategies and create infrastructure. In this specific case, farmers at district, Block and village levels want their *"voice heard and officials of Central and State Governments held accountable"*.

To facilitate better implementation of programs and projects farmers desire to avail most supplies and services at block headquarters or preferably in the village itself. Each Block in the country should become a nodal centre and adequately strengthened to provide all services to farmers, viz. 1. Training to farmers in all respects of scientific methods of cultivation (theoretical with practical demonstration) 2. Sale of approved quality seeds, fertilizers, pesticides, farm tools, equipment and machinery to farmers by authorized dealers; 3. Soil, water, seed, fertilizer, pesticide testing facilities; 4. Supply of spares, services and repair facilities of farm machinery; 5. Storage and warehousing facilities; 6. Agricultural Produce Market Committee; 7. Krishi Vigyan Kendra; 8. Branch of financial institutions supported by technologies.

Each village should have progressive facilities, viz. 1. Seed production by progressive and trained farmers under the close supervision of technical staff and certified by an authorized agency appointed at district level; 2. Agricultural extension staff; 3. Demonstration of technology on selected small farmer's field; 4. Formation of Farmers' clubs and Self-Help-Groups of small, marginal, tenant and women farmers; share croppers and oral lessees; 5. Branch of financial institutions to provide inclusive facilities to farmers and rural households from a single window; 6. Automatic Weather Record Station to help farmers' access weather-based crop insurance scheme; 7. Village Mandi in relatively bigger village Since resources, viz. land, water and energy are limited, scarce, costly and having competing demand for urbanization, industrialization and meeting farming needs, there is a greater need now than before to consider appointing an independent Regulatory and Development Authority manned by professional to look these resources in totality and evolve appropriate legal framework, mechanism and procedure to deal with existing and emerging issues. Farm productivity improvement depends significantly upon the use of inputs, viz. seeds, fertilizers, pesticides, water, farm equipment and machinery, etc. which have necessarily to be of standard quality, available on time and at reasonable price, protecting consumers' rights. This, therefore, calls for putting in place the Regulatory and Development Authority to consider legal framework, mechanism and procedure to ensure that farmers as users invariably are guaranteed to receive farm inputs of these characteristics.

Raising Farmers' Income

There could be various mechanisms to achieve growth in farmers' income within as well as outside the farm sector. Within the farm sector, the increase in incomes can come from greater production per unit area as a result of higher yields, as there are yield gaps of the order of 25%–100% across various crops especially in dryl and regions, and in some high value crops even in irrigated areas. India's average yields in most crops are only half that of China's. The increase can also come from higher cropping intensity, lower costs of production, higher prices, or a combination thereof. But yields are not amenable to sudden change as they depend on the type of technology deployed, which cannot be altered in a very short period of time. Also, high-yielding varieties need irrigation facilities, which cannot be expanded quickly.

Agricultural Growth and Structural Change

There was slow structural change in labour markets at the national level and among developed States. But states such as Kerala, Tamil Nadu, Himachal Pradesh, Punjab and Haryana are on the average of the Lewis Turning Point (LTP) (i.e., situation on economic development) with faster non-farm sector growths, high per capita incomes, urbanisation, higher agricultural labour productivity, and higher wage rates. On the other hand, states with rapid economic growth, such as Gujarat, Andhra Pradesh, West Bengal and Maharashtra, Odisha, Madhya Pradesh, Chhattisgarh, Bihar, have lower wage rates and higher rural poverty. Labour scarcity was visible, as indicated by the negative growth in labour use per hectare and the positive growth in farm mechanisation and agricultural labour productivity. But they also have the potential to pass the Lewis turning point if structural change occurs soon.

Innovative Policy Interventions

The process of conceptualisation, design and implementation of agricultural policies in India is governed bya complex system of institutions and will be over a dozen central ministries at the national level. Although it is the states which have the constitutional mandate for many aspects of agriculture, the central government plays a predominant role by developing national approaches to policy and providing necessary funds for the implementation of programmes at the national and state levels. As the task of governance in the agriculture and allied sectors assumes increasing importance, the need is being felt for a strong arrangement to bring state and central level policy makers together to discuss problems, design solutions and monitor performance. Over the past several decades, agricultural policies have sought to achieve food security, often interpreted in India as self-sufficiency: seeking to ensure that farmers receive remunerative prices, while at the same time safeguarding the interest of consumers by making food available at affordable prices. This approach is resulting in a combination of complex domestic market regulations, import and export trade restrictions, leading to lower prices for farmers that are sometimes below comparable international market levels. Despite large subsidies to fertilisers, power, and irrigation, which offset somewhat the price-depressing effect of market interventions, the overall effect is that policy intervention reduces gross farm incomes.

It is observed that during 2000 and 2016, implementation of policy induced inefficiencies such as the Minimum Support Prices regime offering prices below international prices for several commodities as part of domestic regulations and trade policy measures. Statutory instruments that govern the marketing of agricultural commodities in India include the Essential Commodities Act (ECA) and the Agricultural Produce Market Committee (APMCs) Acts that influence pricing, procuring, stocking, and trading of agricultural commodities. This restricts private sector investment in marketing infrastructure. There are also differences among the states in the status of their respective APMC Acts and in how these Acts are implemented add to the uncertainties in supply chains and drive up transaction costs. The fact is that the combination of market regulations and infrastructure deficiencies has had a price depressing effect in the last 15 years.

Apart from these other inefficiencies in the agricultural market chain also contribute to lower prices to farm produce in India. India's agriculture sector is at the crossroads, exposed, as it is to multiple threats and several opportunities. There is a need for innovative policies to help the country meet increasing food grain requirement and enhance the quality of life to increasing millions of smallholders, overcome severe resource and climate pressures, while generating sustainable productivity growth and creating a modern, efficient and resilient agriculture sector that can contribute to inclusive growth and attract youth to agriculture.

Future policy interventions, thus, need to revisit the MSP regime to reduce the negative effect on farm-gate prices. This will not only help enhance productivity growth but also lead to better resource use, as the extant largely – productive subsidies can be diverted to more and efficient application.

The roadmap for doubling farmers' income by 2022

Doubling Farmers Income (DFI) committee under Shri Ashok Dalwai, CEO, National Rainfed Area Authority

Seven Major Source of Growth:

1 (a).Operating within the agriculture sector

1. Improvement in crop productivity.
2. Improvement in livestock productivity.

3. Resource use efficiency or saving in cost of production.
4. Increase in cropping intensity.
5. Diversification towards high value crops.
6. Improvement in real prices received by farmers.

1 (b) Outside the agriculture sector

1. Shift from farm to non-farm occupations.

The DFI Committee addressed agriculture as a value-led enterprise and suggested empowering farmers with improved market linkages and enabling self-sustainable models as the basis for continued growth in productivity and production income as well as, therefore it should address the basic five primary concerns:

1. Optimal monetisation of farmers' produce.
2. Sustainability of production.
3. Improved resource use efficiency.
4. Re-strengthening of extension and knowledge-based services, and
5. Risk management.

The Committee also highlighted "growth targets" for doubling farmer's real income while improving the ratio between farm and non-farm income- from 60:40 as of now, to 70:30 by 2022-23. It suggested the following strategy:

(a) Adopting a "demand-driven approach" for efficient monetisation of farm produce and to synchronise the production activities in Agriculture & Allied Sectors.

(b) Improving and optimising input delivery mechanism and overall input efficiency [technologies, irrigation methods, mechanisation, Integrated Pest Management (IPM), Integrated Nutrient Management (INM), farm extension services, adaptation to climate change, integrated agri-logistics systems, Integrated Farming Systems Approach, etc.].

(c) Adequate, timely institutional credit support at the individual farmer and cluster levels.

(d) Strengthening linkages with MSMEs (micro, small and medium enterprises) to accelerate growth in both farm as well as non-farm incomes along with employment creation.

The committee deliberate upon specific economic activities and topics that have a durable impact on farmers' income including:

(i) Effective Input Management achieving Resource-Use-Efficiency (RUE) and Total Factor Productivity (TFP) – Water, soil, fertilisers, seeds, labour-farm mechanisation, credit and precision farming, so as to reduce farm losses, while ensuring sustainable and eco-friendly practices.

(ii) Enhancing production through productivity– to achieve and sustain higher production out of less and release land and water resources to diversify into higher value farming for enhanced income.

(iii) Farm Linked Activities to include secondary agriculture that utilises local manpower and biological resource in the vicinity of farms. These can also comprise manufacturing and services activities of KVIC (Khadi and Village Industries Commission) and MSME (Micro, Small and Medium Enterprises) scale, for promoting near-farm and off-farm income generating opportunities as well as to facilitate more of the farm produce to capture more of the market value.

(iv) Agricultural Risk Assessment and Management including drought management, demand & price forecast, weather forecast, management of biotic stress including vertebrate pests, access to credit among farmers for farming operations; providing long term credit, post-production finance to preventing distress sale by farmers, and crop & animal risk management through insurance.

(v) Empowering Farmers through Agricultural Extension, Knowledge Diffusion and Skill Development.

(vi) Research & Development and ICT designed to support the Doubling of Farmers' Income strategy in the short run and help accelerate the pace of income enhancement on a sustainable basis in the long run.

(vii) Structural and Governance Reforms in Agriculture, including building a database of farmers, facilitating farmer & produce mobilisation, an institutional mechanism at district, state & national levels for coordination & convergence, digital monitoring dashboard at the district, state & national levels for seamless &real time monitoring of field delivery, utilising Panchayat Raj Institutions, and farm income measurement as key delivery channels for transparent and inclusive development.

The convergence also called for paying special attention to non-timber forest produce (NTFP) to support tribal farming communities to capture higher value and non-farm incomes there from.

Sustaining Income Growth – Five Pillars

The Committee identified five primary aspects as essential to doubling farmers' incomes, and sustaining a steady income growth in the long run.

(i) Increasing productivity as a route to higher production.

(ii) Reduced cost of production/cultivation.

(iii) Optimal monetisation of the produce.

(iv) Sustainable production technology.

(v) Risk negotiation all along the agricultural value chain.

The underlying theme of the DFI Report is to promote agriculture as an enterprise and farmer as an entrepreneur necessitating the adoption of business principles for positive net returns. Further, the agriculture sector as a profession will become wholesome, when the transition happens from food security to nutrition security for the consumers, extractive production system to sustainable production system for the ecology, and from a mere Green Revolution to move towards a Farmers' Income Revolution or Income Revolution for the farmers.

Strategy for faster and sustainable agricultural growth leading to the doubling of farmers' incomes

Focused efforts on upscaling innovations linked to higher productivity, sustainability and profitability through most appropriate diversified, secondary and specialty agriculture linked to post-harvest management, partially around proper storage, value addition and better access to market – are expected to help in doubling farmers' incomes. In fact, the Green Revolution in itself was an innovation-led initiative around the use of high yielding dwarf wheat and rice varieties that responded favourably to higher inputs such as assured irrigation, chemical fertilizer and pesticides leading to a quantum jump in productivity. Among the cradles of that successful phenomenon were political will combined with strong policy support, committed institutions, improved human resources, well-trained extension workers, the involvement of farmers, and, finally, partnerships at the global level.

Considering the current challenges of low factor productivity, growth decline, depleting natural resources, increasing cost of inputs, higher incidence of diseases and pests, poor returns to farmers and above all the adverse impact of climate change, the task of increasing incomes, especially of 86% farmers who are small and marginal will require technologies by which they can save costs on inputs and derives higher incomes through enhanced productivity and improved linkages to markets. In this regard, the following among other three-pronged strategies need to be pursued seriously:

- Improving productivity and production efficiency

- Diversification - including secondary and specialty agriculture (i.e., fruits and vegetables, tree nuts, dried fruits and horticulture and nursery crops, including floriculture)

- Policy support and better market linkages.

India is expected to achieve the ambitious goal of doubling farm income by 2022. The agriculture sector in India is expected to generate better momentum in the coming years with increased investments in agricultural infrastructure such as irrigation facilities, warehousing and cold storage. Furthermore, the growing use of genetically modified crops will likely improve the yield for Indian farmers. India is expected to be self-sufficient in pulses in the coming few years due to concerted efforts of scientists to get early-maturing varieties of pulses and the increase in minimum support price. The government of India targets to increase the average annual income of a farmer household at current prices to ₹ 219,724 (US$ 3,420.21) by 2022-23 from ₹ 96,703 (US$ 1,505.27) in 2015-16. Going forward, the adoption of food safety and quality assurance mechanisms such as Total Quality Management (TQM) including ISO 9000, ISO 22000, Hazard Analysis and Critical Control Points (HACCP), Good Manufacturing Practices (GMP) and Good Hygienic Practices (GHP) by the food processing industry will offer opportunities for increasing farmers' incomes.

Recommendations

1. The centrality of the smallholder farmer and his linkage with reliable and remunerative markets to be emphasised.

2. Given the high pressure on land, the unmindful splitting of cultivated land – often rendering the holdings un-economical, off-farm and non-farm employment of the farming families is a must.

3. On-farm and rural storage, processing, value addition, marketing, especially through Farmer Producer Organizations (FPOs), should enhance the employment opportunities.

4. Education and training of women to empower them with knowledge and to improve their employment potential will be a lasting approach to overcome the maladies. Moreover, there should not be pay disparity among women and men for an equal job done. This parity alone will enhance agricultural production and income by about 30 per cent.

5. To generate both food and raw material, to meet the requirement of modern society for feed, fibre, fuel and other industrial uses, and in a manner that is sustainable and aims to bring economic growth to farmers.

6. Agriculture has the moral responsibility of meeting the food and nutritional security of the country in consonance with the agro-ecological backdrop.

7. It has to generate gainful employment resulting in income gains to make the farmers more economically secure.

8. It has to generate raw material that will directly support agro-processing of food and non-food products to support secondary agriculture.

9. It has to support the agro-processing industry to produce primary and intermediate goods, which will feed the manufacturing sector.

10. Agricultural practices need to be on a sustainable basis.

11. In an agrarian economy of India's dimension, predominated by unemployed youth and smallholder farmers, entrepreneurship must be a key driver for the agrarian transformation and socio-economic uplift of the majority of youth and farmers who often operate on the edges of the economy in an ever-changing and increasingly complex and competitive global economy. Thus, agriculture should be promoted as an enterprise and farmer as an entrepreneur. The agri-entrepreneurs will greatly strengthen employment security and help harness the huge demographic dividend.

12. Emphasis on start-ups and the role of technology and innovation in India's transformation should be realized on priority basis. The efficacy of recent government initiatives such as Make-in-India, Atal Innovation Mission, Digital India, Skill India, ASPIRE, MUDRA, and Stand-up-India in entrepreneurial development should be analysed and converged for greater impact.

13. The Amul Model, showcased as one of the world's leading Group Entrepreneurship model, should also be applied in other sub-sectors and value-chain systems.

14. The agricultural education system backed up by programs like student READY, RAWE and ARYA should ensure that every agriculture graduate could become an entrepreneur to accelerate equitable economic growth and employment security to build New India.

15. The export of oil meals for animal feed and castor oil has to some extent plugged the import bill of vegetable oil. However, the policy of exporting oil meal may not be desirable in long run as the domestic animal industry is deprived of high value feed to increase milk and meat production. There are also arguments that instead of importing crude or refined oil, why not import oilseeds so that the local crushing industry also prospers and at the same time oil cake as animal feed is also made available. Policies that balance all these factors should be formulated through PPP mode.

16. The partnership can be useful in several aspects of the oil economy such as seed production, forward-backward linkages for processing, value addition, contract research in niche areas, contract farming and joint ventures for higher order derivatives and specialty products. Therefore it is essential to create an enabling environment for private sector participation in oil seeds, pulses sector.

17. About 30-45 per cent of the loss is due to food wastage and the prevention of losses and food waste could feed millions of people. One of the five pillars of Zero Hunger Challenges is "Zero loss or waste of food", seeking change in the mind-set of people to adopt "Save and Grow". Cost-effective on-farm and near-farm processing, value addition along the value-chains, packaging, cold chains, product quality, product safety, and prolonged shelf-life technologies are need of the hour. Let us remember that *"a grain saved is a grain produced"*, and *"unsafe food is no food"*. To achieve this goal smallholder farmers should be directly linked with markets.

18. Greater understanding of market intelligence mechanisms, good trade practices, and legal aspects of the multilateral trade regime and agreements and intellectual property rights is absolutely necessary.

19. New Initiatives of the Government on e-NAM, creating a common Indian Market and investment in establishing additionally 22,000 village markets will directly link the farmers with markets and help them realize remunerative prices for their produces, thus enhancing their incomes and access to adequate nutritive foods. Operation Greens are increasingly focusing on agri-logistics, processing, and professional management.

20. Farmer Producer Organizations (FPOs) and Village Producer Organizations (VPOs) clusters are varyingly incentivized to remuneratively link the farmers with markets directly through the e-NAM platforms. This move has truly created pan-India markets for farmers who could sell their produces at the best prices offered from any corner of the country. Sixteen states and 585 APMC markets are integrated with the e-NAM platform.

21. Enhance our ability to manage natural resources sustainably and with lower negative environmental impacts, and to increase resource efficiency and reduce waste. Adopting Climate Smart Agriculture (CSA), with its triple wins of enhanced productivity, resilience (adaptation) and mitigation, and the associated technological, policy, and investment implications, will facilitate the way to an evergreen future.

22. Accelerate and scale-up actions to build climate smart agriculture and food system at a national and international level; several policies and action platforms are striving to achieve it but these are not aligned. We must integrate these platforms to pursue coherent objectives to insulate people from the veritable vagaries and help achieve comprehensive food and nutrition security and safe health. It requests for increased partnerships and multi-year, large-scale funding of integrated disaster risk reduction and management and climate change adaptation programs that are short-, medium-, and long-term in scope. There is a need for specific

allocations to make it a Pan-India movement to curb the risks to rural prosperity.

23. Emphasize that the availability of water is declining day by day, while on the other hand, there is no dearth of field-tested efficient water conservation and management technologies developed, such as micro-irrigation, sub-surface irrigation and sensor-based water management, and these technologies should effectively be adopted at the grassroots level through creating appropriate institutional mechanisms and policy actions towards all time comprehensive food, water, and nutritional security.

24. It should be emphasized that a series of ecosystem services, crucial for attaining comprehensive food and nutritional security, are provisioned by soil. The centrality of healthy soil for a healthy life should guide all efforts to transform our agriculture-food systems as regards organic farming need to promote strong research backup to develop national standards for organic certification coherent with the international protocols.

25. The technology transfer (extension and training) should be strengthened to ensure the safe use of agrochemicals along the value chain.

26. Science, technology and innovation are key drivers for empowering individuals, societies and nations to meet the challenges, and the richness of this continuum will underpin competitiveness at various levels in this fast expanding knowledge economy era.

27. Encourage "SMART FARMS" with greater technological support to reduce post-harvest losses, strengthen the value chain, and provide markets to increase farmer's income. Agriculture being the main driver of livelihood and economic security of the masses, we need to regularly innovate, invest and support to strengthen agricultural R and D in the country.

28. Agricultural research must be enhanced to bring within its domain cutting-edge technologies like biosensors, genomics, biotechnology, nanotechnology and alternative energy sources.

29. Strengthen innovation drives, leveraging agriculture technology start-ups in India's Innovation, Ecosystem, and Atal Innovation Mission.

30. Precision agriculture exploits all the modern tools and technologies (nanotechnology, genetic engineering, Cloud Computing, etc.), innovations, Artificial Intelligence, etc., leading to economically improved and environmentally sustainable agriculture - an Evergreen Revolution. The role of precision agriculture for the majority of smallholder farmers is manifested through (i) increased land and labour productivity, (ii) intensification, diversification and off-farm

employment, (iii) granting property rights, entitlements and land rights, and (iv) balanced agro-ecological settings compatible with minimum risk. In particular, the roles of Information Technology as the backbone of Precision Agriculture, and of geospatial technologies such as GPS-based soil sampling, drones, robotics, and sensors etc. as framework support are immense. It may be reiterated that "doing the right thing at the right place at the right time" to ensures high efficiency and sustainability are the "watchwords" for Doubling Farmers' Income.

31. Comprehensive and reliable data resources are conducive to augment Artificial Intelligence, as intelligence exhibited by machines, can be applied across disciplines and can bring a paradigm shift in how we see farming today. More specifically, by using artificial intelligence, we can develop smart farming practices to minimize losses along the value chain and maximize outputs and farmers' income. Using artificial intelligence platforms, one can gather a large amount of data from government and public websites or real time monitoring of various data is also possible by using IoT (Internet of Things) and then can be analysed with accuracy to enable the farmers to address all the uncertain issues faced by them in the agriculture sector.

32. Artificial Intelligence (AI) driven solution will not only enable farmers to do more with less, it will also improve quality and become the major driver for providing the digital solutions to uplift the conditions of the farming community while providing yet a new opportunity for business and entrepreneurs by enabling smart farm as a service, and help in identifying most suited farming system.

33. Innovations to promote genomics and gene editing must be priority novel approaches to ensure comprehensive food, nutritional, economic, social and environmental security. Progress in the use of Genetically Modified (GM) crop technology on the other hand has suffered due to the lack of clear policy, and so far BT-cotton remains the only commercialized GM crop.

34. The Government must develop appropriate science-evidenced policies and action plans to transfer proven safe, productive and remunerative technologies, especially genetically enriched quality seed, to meet the social, economic and environmental needs of people, and create the necessary ambience to ensure uninterrupted generation and use of new technologies to help meet our goal of doubling farmers income.

35. A Market Price Stabilisation Fund should be established jointly by Central and State Governments and financial institutions to protect farmers during periods of violent fluctuations in prices; as, for example, in the case of perishable commodities like onion, potato, tomato.

36. An Agriculture Risk Fund should be set up to insulate farmers from risks arising from recurrent droughts and other weather aberrations.

37. An opportunity of Agricultural Insurance Policies should become wider and should also cover health insurance, as envisaged under the Parivar Bima Policy. Seed Companies should provide insurance in the case of GM crops.

38. A well-defined, pro-farmer and pro-resource poor consumer Food Security Policy is an urgent necessity. Food security with home grown food grains can alone eradicate widespread rural poverty and malnutrition, since farming is the backbone of the livelihood security system in rural India. This will enable the Government to remain at the commanding height of the national food security system. Building a food security system and containing price rise with imported foodgrains may sometimes be a short term necessity, but will be a long term disaster to our farmers and farming.

39. Since agriculture is a State subject, every State Government should set up State Farmers' Commission with an eminent farmers as Chairpersons. The membership of the Commission should include all the principal stakeholders in the farming enterprise. Such Commissions should submit an Annual Report to be placed before the respective State Legislature for discussion and decision.

40. There are nearly 1.2 million elected women members in panchayats throughout the country. Today, India has about 253,400 rural local bodies at the village level Gram Panchayats, and 6613 intermediary or block-level Panchayats, and 630 district level panchayats. There are about 3 million elected representatives of these panchayats, out of which 1.3 million are women. Thus, 70 per cent of India's population is covered through these local governance institutions. They can play a pivotal role in improving the quality of life in villages and in areas like sanitation, drinking water, child care, early childhood education and nutrition security. They should be empowered to take up such leadership roles in rural transformation through appropriate training and capacity building opportunities. An earmarked Gram Mahila Fund should be available to them for meeting gender-specific needs.

41. Measure to manage is most crucial in India as there exists a huge technology transfer gap, and several important targets stand unmet. It is more important to emphasise on monitoring progress towards achieving a New India free from hunger and malnutrition within the framework of SDGs is essential to drive the necessary structural and policy reforms - "leaving no one behind".

4.2.1 Towards Oilseeds and Pulses Self

Way Forward

The agriculture sector is in dire need of transformation in terms of not only productivity enhancement but also for creating a robust eco-system right from the production to the consumption stage. This is critical for India's overall food security, better management of food inflation, as well as to meet the government's target of doubling farm incomes, while dealing with climate change.

Pathways for Doubling Farmers' Income

The scope for Doubling Farmers' Income (DFI) by 2022 exists with the following compulsions viz., increasing the physical outcome from agriculture, diversification of enterprises, pricing mechanism, adoption of risk management tools, wage rate and salaries for farm labour as well as different sources of non-farm income. A pathway has been formulated which is applicable to all commodities and sectors in agriculture. However, the implementation strategies should be region and need specific targeting different components in agriculture. The framework integrating technology, extension, institutions and policies to double the farmers' income.

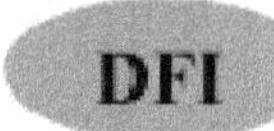

Figure 4.1 Pathways for Doubling Farmers' Income (DFI) by 2022

Technology: Technology is the outcome of science as it enables increased output with the same input or realise the same output with reduced input. Productivity in a majority of the commodities has struck a plateau demanding barrier breaking intervention through cutting-edge sciences. The diversification of farm activities towards high value crops and enterprises can more than quadruple income from the same piece of land. Another augmenting factor for doubling farm income is through integrating crop production with allied and subsidiary enterprises in uncultivated lands.

Extension: Given the technology and resources, the output level can be enhanced by consolidating the existing potentials by bridging the yield gaps between agronomic potentials achieved in research and extension farms and the actual yields obtained in average farmer's fields. Around 30 per cent of the physical output can be increased which will facilitate to increase the farmers' income without any additional investment, but just by adopting the recommended package of practices across different sub-sectors in agriculture.

Institutions: Irrigation is the best insurance against drought. The thrust of the state is to ensure *'Per drop more crop'* through accelerated irrigation schemes supplemented by the massive promotion of micro irrigation techniques for maximum coverage of irrigated crop area. As agriculture in India is a gamble with the monsoon, and the economic pursuit of the enterprise is worth if and only when adequate risk management options are available. There is a need for innovative marketing that will link farmers to productivity tools and insurance.

The Pradhan Mantri Krishi Sinchai Yojana (PMKSY) promotes "more crop per drop" as a national mission to improve farm productivity with a focus on massive expansion of micro irrigation at the farm level with a potential of irrigating over 7.6 million ha. These will reduce the monsoon dependency insulating from shocks and improve productivity.

Prime Minister's Crop Insurance Scheme is based on crop income than the cost of cultivation/credit and thus addresses volatility in farm income besides natural calamities and post-harvest losses, unlike earlier schemes that covered only loss of yield or investment or capital besides. It is planned to enhance the coverage of the scheme to more than 50 per cent of the cropped area by next year. This coupled with land leasing laws will enable farmers including tenants to absorb risk and invest more into better farm techniques and crops.

The recently initiated National Agricultural Marketing brings more than 500 markets on a single e-platform integrating markets sans middlemen and enabling farmers to bid their products to sell anywhere in India. But the integration will have operational meaning for the farmers only when the Agricultural Produce Market Committee (APMC) Act and Essential Commodities Act (ECA) are either revamped or done away with. This is very much evident in the difference in realization of the price potential between livestock products not covered under these acts and the fruits and vegetables covered under the APMC. Further, it will not suffice if commodities are delisted under APMC unless alternate marketing platforms are made available with the attendant, scientific storage, processing and transport facilities where private can play a vital role.

Policy: Policy on agricultural commodities export and stocking have more often than not acted against the farmers' interest preventing them from partaking in the benefits of the free market. Whenever prices shoot up, banning or restricting the export or limiting stocking of the commodity have been resorted to more as a panic reaction than a planned strategic measure. By these measures, India has gained notoriety for being an unreliable supplier for the international market denying its farmers their due benefits from participating in the international trade. Besides, the uncertainties regarding stocking norms, no investment worth its while can ever happen on scientific storage, processing, value addition and contract farming.

Unless agricultural commodities are freed from the ambit of Essential Commodities Act or the Act is repealed altogether, income from and investment in agricultural marketing is not possible. When there is a spike in prices, the difference between 22 modal market prices and MSP may be made up through direct transfer of deficiency payment to the farmer. Terms of trade is another important policy tool to enhance farmers' income by tweaking the prices of farm products compared to their non-agricultural counterparts but without stoking food inflation. Inflation in agricultural prices also leads to an enhanced real farm income if prices received by farmers increase at a faster rate in comparison to the prices paid by them. For the past few years, the Wholesale Price Index (WPI) based inflation of non-agricultural prices is declining, whereas that of the agricultural prices has been increasing by about 5.5 per cent, implying a 5.5 per cent growth in real farm income. If technology and factor prices could result in per-unit cost savings, farmers' income would rise at a much higher rate than the rate of increase in output.

Strategies for Doubling Income: Income is the most relevant measure to assess the farmers' welfare and agriculture transformation. Even today, the highest returns on investment on per unit basis are from agriculture. What is lacking is the scale unlike corporate investment. Certainly, returns from cultivation alone will not help to achieve the set target of DFI. It has to be supplemented, in fact, to a larger extent by livestock and other non-farm activities supported with policy intervention at all levels.

Table 4.1 Potential strategies to enhance annual farm household income

Parameter	Strategy
Science and Technology	• Adoption of improved varieties/ breeds/ strains for additional income • e-Grid of all-weather stations for providing location specific weather information • Nutrients sale based on Soil health card programme • Adoption of micro irrigation system to improve the water use efficiency

Table 4.1 *Contd…*

Parameter	Strategy
Extension	• Bridging the gaps between achievable (FLD) and potential yields • More number of effective cluster demonstrations to bridge the information gap • Village adoption to transfer the technologies developed by the research organizations • Upscaling and out scaling of technologies through field days, exhibition and other activities
Policies	• Rationalizing the subsidy on energy use • Enrolling more number of marginal and small holders under the crop insurance scheme. • Integrating all central and state subsidies in agriculture • Formation of Crop Planning Department at national and state level. • Additional investment in agricultural R&D to pave the path for innovation. • Setting up more organic food certification agencies. • Policy for setting up of FPO for block-level seed production. • Integrated land-use policy particularly for water • Breaking of crop monotony and focus on diversification
Institutions	• Setting up of Agribusiness Centres at the district level • Transparency and simplified procedures in electronic trading • Developing a comprehensive framework for community / corporate farming. • More emphasis on e-learning in regional languages • Value chain development for primary commodities

Source: Author Compilation

Policy Interventions

To make agriculture both remunerative and attractive as a profession, and especially to double the farmers' income, an action plan for implementing the three-pronged strategy proposed above is described here: Policy Interventions, a 'National Mission on Farmer First', with an annual allocation of INR 10,000 crores to begin with and by merging/clubbing of various central schemes as well as through some new initiatives to empower farmers needs to be initiated soon. This will help in catalysing the activities/ programmes specifically designed for scaling innovations that will increase farmers' income and have a direct impact on smallholder farmers through the adoption of a three-pronged strategy.

Needed regulatory reforms in the existing Acts especially pertaining to the land, water, seed, fertilizer, energy and market, etc. must be brought about as a matter of national priority by the central government. Also, to have in place an effective coordination and convergence mechanism for various schemes, programmes and activities by different Ministries would help in achieving desired outcomes much faster. For this, a high-level Inter-Ministerial Committee to be chaired by the Prime Minister and co-chaired by the Vice-Chairman, NITI Aayog and Agriculture Minister will help ensure effective monitoring of the outcomes of various programmes aiming at 'Farmer First'. Also, this coordination committee be assisted by a standing Advisory Panel of Agricultural Experts.

Remunerative minimum support price (MSP) for most of the commodities needs to be fixed and announced well in advance of planting season by the Ministry of Agriculture and Farmers Welfare (MOA&FW) with assurance for either procurement or compensation directly to the producers for the prevailing price difference in the market so that farmer is not a looser. Also, the reforms in methodology for fixing MSP by the Commission for Agricultural Costs and Prices (CACP), is essentially needed, for which a High Level External Review Committee of Experts be established immediately.

For accelerating agricultural growth, needed incentives and rewards must be put in place quickly to attract youth (including women) to diversified, secondary and specialty agriculture as individual producers, SHGs, Cooperatives, Farmers Producer Organizations/Companies or as knowledge/ service providers. In the process, farmer-led innovations must be scaled out through required validation, refinement and incentives in the form of credit at low-interest rates (not >4%), bank support for required commercialization, insurance to avoid any initial risks, practically no or very low tax on rural-based value additions and marketing of produce/value-added products. Incentives to innovators/ entrepreneurs could be in the form of state/national recognitions and awards.

Right policy support for the accelerated role of the private sector will certainly change the game much faster. Hence, enabling the environment to embrace the private sector is the most critical need which should be given due importance by the Government. In this context, support for hybrid seed production; fabrication of equipment /implements/ tool for scaling conservation agriculture and small farm mechanization; micro-irrigation (drip and sprinkler), protected cultivation, including fertigation; agro-processing and value addition; fertilizers, including customized bio-fertilizers; pesticides, including bio-pesticides, etc. would help accelerate agricultural growth.

The Ministry of Agriculture needs to undergo a self-imposed "one-time-catch-up" exercise to remain au fait with the rapidly changing external environment. So should be the players in the value chain. There is also no effort at macro-management in important areas of the Centre's basic functions - such as credit, insurance, research and Exim Policy (which is currently, at best, stop-go and knee-jerk in nature).

Many important and urgent steps need to be taken to ensure the rapid and sustainable growth of the agriculture and allied sectors in the future, and for farm-incomes to increase to expected levels. It must be appreciated that the process of decision-making in agriculture and related sectors occurs in a complex and vast administrative space. Therefore the requisite speed and accuracy cannot obtain unless the highest levels in government at the Centre and State levels directly participate in the process and lead from the front. To that end, it is necessary to constitute a Cabinet Committee presided over by the Prime Minister at the national level in which Ministers responsible for all the departments concerned with agriculture and the allied sectors are Members. The Committee should be serviced by the National Institute of Agricultural Extension Management (MANAGE) through a trans-disciplinary team (whose functioning is informed by the process of chemistry among its members rather than mere physics). Disciplines such as Research, Extension, Meteorology, Financial Services including Banking, Insurance, Agronomy, Animal Husbandry, Storage, Processing and Marketing. Similar mechanisms should obtain at the state and district levels with the Chief Ministers and the District Magistrates/Collectors as the heads – serviced by SAMETIs and ATMAs respectively. At all levels an online and real time scanning should be undertaken with a view to spotting opportunities and threats, adding value thereto and dispatching the resultant messages to appropriate destinations in the country.

The centre and the states should function on their own in there in certain spheres. For instance, the centre alone will need to deal with areas such as international cooperation and agriculture, international trade, financial services such as banking and insurance. The centre and states should act in a complementary manner in areas such as the setting up of and maintaining of infrastructure such as ports, railways, national highways, cold storage chains and the like, and management of the National Research System in agriculture and allied sectors. Similarly, the states should deal exclusively with areas such as extension, the designing, preparation, and implementation of district plans to be financed by the funds available at the state level, etc. In all other areas the Centre and the states should supplement the activities of one another.

In place of the extant arrangements, the centre and the states should jointly enter into Memoranda of Understanding (MoUs) which promise pre-set, and quantified deliverable outputs in terms of production, productivity, and other thematic imperatives such as soil health, centre-staging the role of women, NRM, environment-friendly practices, robust responses to the challenges of climate change, etc. The sum of the outputs and outcomes of all the MoUs in the country should, ideally, translate into national objectives, both in terms of quantities and thematic imperatives.

Research and Development

- Besides the focus on productivity and production growth, we now need increased research and development emphasis on post-production, value addition, and market linkages (both domestic and foreign).

- There is an urgent need to improve the empowerment of targeted smallholder farmers and ensure the delivery of last-mile services. Hence, the technology dissemination related programmes will have to be tailored and reoriented according to present day needs. In fact, a paradigm shift from public to private innovation extension system is the need of the hour to provide much needed knowledge, quality inputs and much needed custom hire services at the farmer's doorstep.

- It needs to be ensured that smallholder farmers, especially the youth including female farmers, get their entitlements and are not side lined.

- Identification of agencies/institutions responsible for taking specific actions at the local, State and Central level and their effective coordination will be very helpful. Also, an independent monitoring and evaluation process for the much-needed impact will be extremely useful.

Capacity Development

- Knowledge sharing and capacity development (especially women and youth) need to be considered a top priority to bridge the yield gaps, achieve diversification, scaling innovations that can save on production costs and help in the rational use of natural resources, ensure value addition and link the farmers to market.

- Greater emphasis must be given henceforth on skill (on the farm as well as off-farm activities) development at all levels. This will greatly help the farmers especially the smallholders to raise their income.

Financial Support

- There is an urgent need to triple annual budget allocation for the Indian Council of Agricultural Research (ICAR), an apex AR4D organization with a proven track record, in order to continue meeting emerging challenges while providing national public goods for the betterment of farmers as well as Indian agriculture.

- Capital investment in agriculture for much required infrastructure in the States, which were left behind during the Green Revolution period (especially the eastern region), must immediately be enhanced (at least to a minimum level of 15-20% from the present level.

Improved Agricultural Productivity

- Consolidate land by forming collectives/ farmer organisations/ co-operatives (on the lines of AMUL). This would help in achieving economies of scale for operation and provide farmers the wherewithal to tap into specific technological innovations and reach out to a wider consumer base. Another option which could be considered is the long-term leasing of farm landholdings without alienating land ownership.

- Enhance the level of mechanisation through the adoption of the Custom Hiring Model. Under this model, a group of farmers like Farmer Producer Organisations (FPO) and cooperatives purchase high-cost agricultural machinery which can be used by all farmers in return for a fixed payment. Local manufacturing of small and low-end agri machineries should be supported under the 'Make in India' initiative to restrain the cost of production and price of machineries to improve affordability. Adequate service centres should be established in rural and remote areas to offer after-sales service. Financial assistance, loans, and subsidies for mechanisation should be continued.

- Encourage adoption of digital technology such as remote sensing (via satellites), use of drones, GIS, etc., for sustainable farming. Unmanned aerial vehicles with powerful cameras/drones can be used to assess the fertilisation status of crops. The sensor network can be developed using smart technology for continuous monitoring of farms and specifically the use of virtual fence technologies in the livestock sector where the movement of livestock can be monitored with the use of remote sensing signals or sensors. It would reduce the turnaround time to react or respond to different situations. Uptake of these technologies would enable farmers to earn higher yields, reduce chances of crop failures and earn higher and largely uniform returns. However, since the associated costs may not be affordable for individual farmers, cooperatives could be formed to take up such facilities.

- Provide real time information regarding weather patterns to farmers using digital and mobile technology to help them take better and timely farming-related decisions and lower incidents of crop failures.

- Increase agriculture R&D to develop climate resilient and short-duration high-yielding crop varieties which are helpful in raising cropping intensity.

- Encourage adoption of modern farming techniques such as Organic Farming, Low External Input Sustainable Agriculture (LEISA) and Precision Farming (an information and technology-based farm management system to identify, analyse, and manage variability within fields for optimum sustainability, profitability and protection of the land resource).

- Encourage adoption of smart irrigation techniques such as micro-irrigation or drip irrigation system which holds the potential to double the yield using

50 per cent of the water required with traditional methods and helps in increasing efficiency of other farm inputs such as fertilisers, pesticides, and labour.

Reduce post-harvest losses through the development of efficient cold chain infrastructure Strengthen and improve cold chain infrastructure in the country: Set up warehouses and cold storage facilities in each district to ensure that farmers have adequate access to such facilities. These facilities should be affordable, reliable and sustainable. The quality of existing facilities should be improved with the latest technology such as hermetic storage structures which can prevent storage losses, while maintaining the freshness of fruits and vegetables and seed viability and retain quality. Similarly, modern cold chains enabled with climate control technology can help in reducing the rate of metabolism in harvested fruits and vegetables and extend the shelf life of the produce. The use of Artificial Intelligence (AI) can be explored which allows monitoring of a large basket of agri-commodities and micromanage their storage environment remotely, with the use of sophisticated sensors and other equipment.

Implement a PPP based model for enhancing investments in agri supply chain including development and modernization of storage facilities as well as grading and standardization, quality certification, etc.

Improve price realisation through providing farmers' access to market and improving marketing Infrastructure.

Encourage states to adopt the model Agricultural Produce and Livestock Marketing (APLM) Act that proposes a liberalised marketing environment for agri-commodities and livestock. Specific incentives can be given to States adopting the model such as allocation of higher funds under Central Government schemes for agri-development.

Government Policy, Research and Development (R&D), Extension and Good Governance

NABARD, by its Mandate, is ideally suited for initiating appropriate policies for sustainable Agriculture and Farmers' incomes through its Research and Training strategies for which it should put in place strong core strength of techno-economic and credit specialists.

The core areas requiring immediate action are:

1. Deepen Agricultural Strategy/Sector Analysis in NABARD's Annual State-level Potential-linked Credit Plans.

2. Expand Term Finance/Project Finance for Value Chain Development in Agriculture: NABARD, could select one or two states as pilots to explore how Annual Potential-linked Credit Plans could be strengthened to include agricultural sector analysis in greater depth, and explicitly include

the potential for project-based term finance for value chain development in agriculture. NABARD could also associate NABCONS to work on project-based finance proposals initially for a few pilot states and explore PPPs potentially using RIDF funding in conjunction with private sector investments.(The present RIDF funding needs to be revisited to re-align.

3. Address Risk Management Issues in Agriculture and Rural Finance: It is by now widely recognized that India's approach to risk management in agriculture can at best be described as "partial" or "inadequate" and must be strengthened in terms of both policy and content. Increasing number of farmer suicides in states like Maharashtra and Telangana is reinforcing the urgency to comprehensively address this issue. NABARD may therefore consider appointing a high-level working group to review and update the status of existing agricultural risk management framework/s and recommend how these can be improved over time. In this connection, refer to OECD's 2011 publication titled "Managing Risk in Agriculture: Policy Assessment and Design". This publication elaborately lays down risk management principles and guidelines for policy design in agriculture. It deals with risk exposure and risk management at the farm level, as well as exogenous risks and price variability, supplemented by examples of risk management frameworks used in Australia, Canada, Netherlands, New Zealand, and Spain. While India's agricultural environment and risk management have their unique characteristics derived as these are from both real and political economy, a way has to be found to manage agricultural risks (including natural disasters) without necessarily resorting to loan rescheduling and loan waivers or write-offs.

4. Potential Role of NABARD in Climate Change Adaptation Projects: NABARD has since been recognized as the national accredited agency to channel international and national funds for climate change mitigation and adaptation for eligible projects. The landscape approach involves working on complex landscapes in an integrated, holistic manner incorporating different types of land uses (agriculture, forestry, grasslands, wastelands, etc.) in an integrated, holistic manner, using a single management process.

5. R&D- Innovations and Extension: While ICAR should re-design, deepen and widen its macro research in line with the emerging priorities, NABARD may work jointly with ICAR to register native innovations at farmers' level and build up research).It is already an existing component in NABARD's micro planning exercise. Awareness and adoption in relevant local areas (Reverse Research- Micro Interestingly, ICAR has plans in place to scientifically validate, scale-up and propagate the innovations of progressive farmers. Emphasizing the need to encourage the use of technology in the farm sector, ICAR has already created a

linkage with over a hundred Start-Ups. The joint efforts of the carefully planned R&D programs, with the financial and manpower resources at their command, can take the farming to an elevated platform, paving the way for better returns and higher incomes to farmers.

6. Principles of Governance in the Management of "Agriculture", the Prime Sector in India

Observing meticulously the "Principles of Governance" has assumed great importance in the conduct of any activity or business in the post-economic reforms era and consequent global competition.

The three major components of Good Governance are accountability, transparency and prudence. In the Agriculture sector, policies are often compromised with practices by those at the helm of affairs. The policy makers are well advised to reset their respective roles to bring them in consonance with the principles of Good Governance.

4.3 The Farm to Fork Approach

The Indian farming market was worth INR 18,367 Billion in 2019. The farming sector constitutes one of the most important areas of the Indian economy. India currently represents the world's largest producer of fresh fruits and vegetables, major spices, selected fibrous crops such as jute, several staples such as millets and castor oil seed and the second largest producer of wheat and rice, the world's major food staples. Currently, India ranks within the world's five largest producers of over 80% of agricultural items, including cash crops such as coffee and cotton. Crop yields have also increased significantly over the last several decades. Factors such as farm mechanization, increasing usage of fertilizers, improving irrigation techniques, better seeds and easy availability of credit can be regarded as the major drivers of the Indian farming industry.

The Indian food sector has emerged as high-growth and high-profit sector due to its immense potential for value addition, particularly within the food processing industry. Food processors, importers, wholesalers, retailers, and food service operators are all part of a developing agribusiness sector. Agri-value chain is needed to prevent huge wastage of agricultural produce which saves "billions of dollars" loss to economy, ensure that share of the farmer in consumers wallet can increase, which is merely 25-30 per cent as compared to the western market where it goes as high as 50 to 75 per cent. Facilitate demand for Minimum Support Price (MSP) regime which will automatically go away, and farmers can expect a rightful price for the produce. Ensure optimal management of natural resource and make India becomes "Global Hub for Food Industry".

Food Value Chain

The expression "farm-to-fork", is often used to describe food value chains. A food product moves from upstream in the chain, where farmers grow and harvest it, towards the market – through intermediaries including producers' organisations, processors, transporters, wholesalers and retailers on the upstream level of consumers.

First we must fix missing links in the farm-to-fork value chain and also reduce food wastage in the agricultural and food-processing industry. There is a "substantial" difference in what the farmers get for their produce, and what the consumers pay both within the country and internationally is substantial. There is huge wastage not only in farm fields but also in the food processing industry.

Adopt monetization of produce as the basis for maximising the value capture for the farmers. Maximise monetization possibilities by upgrading and harmonising the agri-logistics (storage and transportation), agro-processing and marketing. Adopt market architecture and alternate wholesale markets (APMCs in the private and public sector), and the export market. Promote agricultural value system as a link between farms and markets.

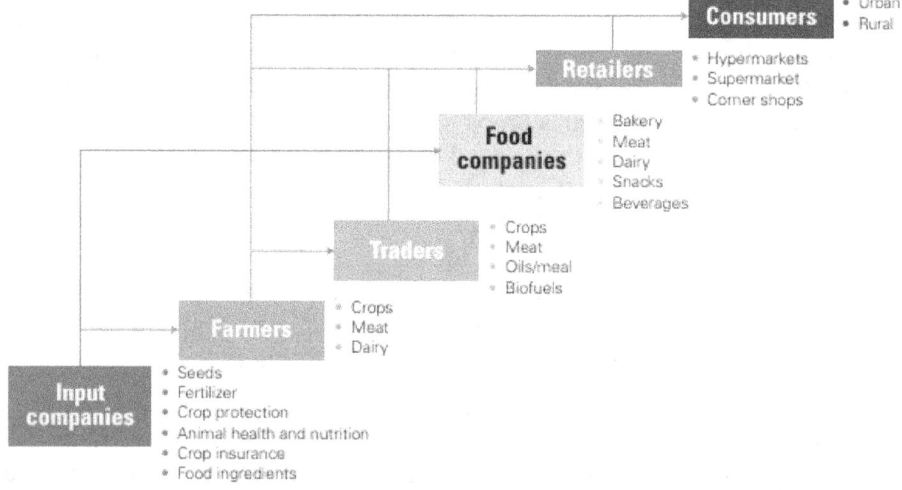

Figure 4.2 The Agricultural Food Value Chain

Conclusion

In India, while farmers are the major producers, they also constitute the largest proportion of consumers. Hence, improving small farm production and productivity, as a major development strategy, can make a significant contribution towards the elimination of hunger and poverty, provided farming is made efficient and remunerative. The experience of countries that have succeeded in reducing hunger and malnutrition shows that growth originating in agriculture, through smallholder farmers, is at least twice as effective in benefiting the poorest as growth from non-agriculture sectors. The World Development Report of the World Bank (World Bank, 2008) has clearly emphasized that: 'Using agriculture as the basis for economic growth in agriculture-based countries requires a productivity revolution in smallholder farming'. As stated earlier, higher productivity requires a higher investment in agriculture and agricultural research - a fact that needs to be heeded by the policy makers to make sure that 1.0% of agricultural GDP is invested in AR4D, as against the present level of just 0.4%. Hence, a three-fold increase in resource allocation for the national agricultural research system (NARS) be considered a prerequisite to double the farmers' income.

It is also a fact that India will remain predominantly an agricultural country during most of the 21st century. Therefore, we must have both the vision and national strategy for shaping the destiny of agriculture by making it highly productive, efficient and economically attractive for the smallholder farming community. The target of doubling farmers' income by 2022, though apparently not easy yet a very laudable goal, augurs well of Government's intention to help farmers. It is also clear that if concerted efforts, as per the suggested action plan, are made in a Mission Mode, chances of making agriculture an engine of national economic growth and for smallholder farmers a respectable profession are indeed much brighter.

A persistent low level of farmers' income can also cause a serious adverse effect on the future of agriculture in the country. To secure the future of agriculture and to improve the livelihoods of half of India's population, adequate attention needs to be given to improve the welfare of farmers and raise agricultural income. Achieving this goal will reduce the persistent disparity between farm and non-farm income, alleviate agrarian distress, promote inclusive growth and infuse dynamism in the agriculture sector. Respectable incomes in the farm sector will also attract the youth towards the farming profession and ease the pressure on non-farm jobs, which are not growing as per the expectations. Doubling farmers' income by 2022 is quite challenging but it is needed and is attainable. The three-pronged strategy focused on 1. Development initiatives, 2. Technology and 3. Policy reforms in agriculture is needed to double farmers' income.

Research institutes should come with technological breakthroughs for shifting production frontiers and raising efficiency in the use of inputs. Evidence is growing about the scope of agronomic practices like precision farming to raise the production and income of farmers substantially. Similarly, modern machinery such as laser land leveller, precision seeder and planter, and practices like SRI (system of rice intensification), direct seeded rice, zero tillage, raised bed plantation and ridge plantation allow technically highly efficient farming. However, these technologies developed by the public sector have very poor marketability. They require a strong extension for the adoption by farmers.

Research and Development (R&D) institutions should also include in their package, grassroots level innovations and traditional practices which are resilient, sustainable and income enhancing. Indian Council of Agriculture Research (ICAR) and State Agriculture Universities (SAUs) should develop models of farming systems for different types of socioeconomic and bio-physical settings combining all their technologies in a package with a focus on farm income. This would involve combining technology and best practices covering production, protection and post-harvest value addition for each sub systems with other sub systems like crop sequences, crop mix, livestock, horticulture, forestry. Such a shift requires an interdisciplinary approach to develop knowledge of all disciplines.

About one-third of the increase in farmers' income is easily attainable through better price realization, efficient post-harvest management, competitive value chains and adoption of allied activities. This requires comprehensive reforms in market, land lease and raising of trees on private land. Agriculture has suffered due to the absence of modern capital and modern knowledge.

There is a need to liberalise agriculture to attract responsible private investments in production and market. Similarly, Farmer Producer Organisations (FPOs) and Farmer Producer Companies (FPCs) can play a big role in promoting small farm businesses. Ensuring MSP alone for farm produce through a competitive market or government intervention will result in a sizeable increase in farmers' income in many states. Most of the development initiatives and policies for agriculture are implemented by the States. States invest much more than the outlay by the Centre on many development activities, like irrigation. Progress of various reforms related to market and land lease are also State subjects. Therefore, it is essential to mobilise States and UTs to own and achieve the goal of doubling farmers' income. If concerted and well-coordinated efforts are made by the Centre and all the States and UTs, the Country can achieve the goal of doubling farmers' income by the year 2022.

REFERENCES

1. Birthal, P.S., Negi, D.S. and Roy, D. (2017), Enhancing Farmers' Income: Who to Target and How?, Policy Paper No.30, ICAR-National Institute of Agricultural Economics and Policy Research, New Delhi, xiv+38.

2. Tracking productivity: The GAP IndexTM

3. The future of food and agriculture – Trends and challenges, FAO

4. The State of Agricultural Commodity Markets: FAO

5. Food and agriculture policy decisions, FAO

6. Pocket Book of Agricultural Statistics, 2018, Directorate of Economics & Statistics

7. Press Information Bureau, GOI

8. Paddock, William and Paul (1968) Famine 1975, Wiedenfeld and Nicolson, London.

9. Pearson, Lester B. (1969) Partners in Development: Report of the Commission on International Development, Praeger Publishers, Inc., Pall Mall Press, London.

10. M.S. Swaminathan (2003) -The Face of Indian Agriculture, RICAREA.

11. Rathore, LS, Attri SD and Jaswal AK. (2013). State Level Climate Change Trends in India: Meteorological Monograph No. ESSO/IMD/EMRC/02/2013. India Meteorological Department.

12. Rathore, LS, Attri SD and Jaswal AK. (2013). State Level Climate Change Trends in India: Meteorological Monograph No. ESSO/IMD/EMRC/02/2013. India Meteorological Department.

13. N. Chattopadhyay, S. Sunitha Devi, Gracy John, and V. R. Choudhari. (2016). Occurrence of hail storms and strategies to minimize its effect on crops. MAUSAM, 68, 1 (January 2017), 75-92.

14. Botzen, W.J. Wouter& Bouwer, Laurens & van den Bergh, Jeroen. (2010). Climate change and hailstorm damage: Empirical evidence and implications for agriculture and insurance. Resource and Energy Economics. 32. 341-362.

15. American Meteorological Society Glossary (n.d.). Thunderstorms. Retrieved February 2019 from http://glossary.ametsoc.org/wiki/Thunderstorm.

16. Kamaljit Ray, B. A. M Kannan, Pradeep Sharma, Bikram Sen and A. H. Warsi, India Meteorological Department. Severe Thunderstorm Activities over India during SAARC STORM, Project 2014-15: Study Based on Radar.

17. Basu, S., Bieniek, P.A. &Deoras, A. Asia-Pacific J Atmos Sci (2017) 53: 75. https://doi.org/10.1007/s13143-017-0006-7.

18. Jamwal, N. (2018, May 24). What's behind the intense storms battering North West India? Experts divided, call for more research Scroll.

19. Jamwal, N. (2018, April 20). India's Met department has forecast normal monsoon this summer – but don't start celebrating yet Scroll.

20. Singh, OP & Khan, Tariq & Rahman, S. (2001). Has the frequency of intense tropical cyclones increased in the north Indian Ocean? Current science. 80. 575-580.

21. Maurya,L. and Goswami S. (2018). Extreme is the new normal. Down to Earth.

22. United Nations Office for Disaster Risk Reduction (UNISDR), 2015a, Making Development Sustainable: The Future of Disaster Risk Management. Global Assessment Report on Disaster Risk Reduction.

23. https://icar.org.in/files/vision-2020.pdf

24. www.goldmansachs.com/our-thinlcing/technology_driving_innovation/drones/

25. http://commerce.gov.in/writereaddata/uploadedfile/MOC_636571617294 118126_Draft_Agr_Export_ Policy.pdf

26. http://pib.nic.in/newsite/PrintRelease.aspx?relid=184369

27. http://commerce.gov.in/writereaddata/uploadedfile/MOC_636571617294 118126_Draft_Agr_Export_Policy.pdf

28. http://apeda.gov.in/apedawebsite/six_head_product/cereal.htm

29. http://apeda.gov.in/apedawebsite/six_head_product/FFV.htm

30. https://www.ibef.org/industry/agriculture-india.aspx

31. https://www.ibef.org/industry/agriculture-india.aspx

32. http://apeda.gov.in/apedawebsite/organic/Organic_Products.htm

33. http://www.apeda.gov.in/apedawebsite/six_head_product/PFV_OPF.htm

34. http://www.apeda.gov.in/apedawebsite/six_head_product/PFV_OPF.htm

35. Details available at http://mowr.gov.in/sites/default/files/Guidelines_ for_improving_water_use_efficiency_1.pdf, last accessed on 26 June 2019

36. Details available at http://www.worldbank.org/en/news/feature /2011/09/29/india-water, last accessed on 26 June 2019

37. Details available at http://mowr.gov.in/sites/default/files /NWP2012Eng6495132651_1.pdf, last accessed on 26 June 2019

38. GoI. Government of India 2011: Agriculture & Cooperation.

39. NAPCC (2008) National Action Plan on Climate Change. Prime Minister's Council on Climate Change, Government of India.

40. MoA (2002), Basic Animal Husbandry Statistics 2002. Ministry of Agriculture, Department of Animal Husbandry & Dairying, Government of India, New Delhi.

41. Jamwal, N. (2019, January 30). Extreme weather events have become the new normal. DNA India.

42. Rathore, LS, Attri SD and Jaswal AK. (2013). State Level Climate Change Trends in India: Meteorological Monograph No. ESSO/IMD/EMRC/02/2013. India Meteorological Department.

43. N. Chattopadhyay, S. Sunitha Devi, Gracy John, and V. R. Choudhari. (2016). Occurrence of hail storms and strategies to minimize its effect on crops. MAUSAM, 68, 1 (January 2017), 75-92.

44. Botzen, W.J. Wouter& Bouwer, Laurens & van den Bergh, Jeroen. (2010). Climate change and hailstorm damage: Empirical evidence and implications for agriculture and insurance. Resource and Energy Economics. 32. 341-362.

45. American Meteorological Society Glossary (n.d.). Thunderstorms. Retrieved February 2019 from http://glossary.ametsoc.org/wiki/ Thunderstorm.

46. Kamaljit Ray, B. A. M Kannan, Pradeep Sharma, Bikram Sen and A. H. Warsi, India Meteorological Department. Severe Thunderstorm Activities over India during SAARC STORM, Project 2014-15: Study Based on Radar.

47. Basu, S., Bieniek, P.A. &Deoras, A. Asia-Pacific J Atmos Sci (2017) 53: 75. https://doi.org/10.1007/s13143-017-0006-7.

48. Jamwal, N. (2018, May 24). What's behind the intense storms battering North West India? Experts divided, call for more research Scroll.

49. Jamwal, N. (2018, April 20). India's Met department has forecast normal monsoon this summer – but don't start celebrating yet Scroll.

50. Singh, OP & Khan, Tariq & Rahman, S. (2001). Has the frequency of intense tropical cyclones increased in the north Indian Ocean? Current science. 80. 575-580.

51. Maurya,L. and Goswami S. (2018). Extreme is the new normal. Down to Earth.

52. United Nations Office for Disaster Risk Reduction (UNISDR), 2015a, Making Development Sustainable: The Future of Disaster Risk Management. Global Assessment Report on Disaster Risk Reduction.

53. https://icar.org.in/files/vision-2020.pdf

54. www.goldmansachs.com/our-thinlcing/technology_driving_innovation/drones/

55. http://commerce.gov.in/writereaddata/uploadedfile/MOC_636571617294118126_Draft_Agr_Export_ Policy.pdf

56. http://pib.nic.in/newsite/PrintRelease.aspx?relid=184369

57. http://commerce.gov.in/writereaddata/uploadedfile/MOC_636571617294118126_Draft_Agr_Export_Policy.pdf

58. http://apeda.gov.in/apedawebsite/six_head_product/cereal.htm

59. http://apeda.gov.in/apedawebsite/six_head_product/FFV.htm

60. https://www.ibef.org/industry/agriculture-india.aspx

61. https://www.ibef.org/industry/agriculture-india.aspx

62. http://apeda.gov.in/apedawebsite/organic/Organic_Products.htm

63. http://www.apeda.gov.in/apedawebsite/six_head_product/PFV_OPF.htm

64. http://www.apeda.gov.in/apedawebsite/six_head_product/PFV_OPF.htm

65. On secondary agriculture, see Chengappa (2016)

66. NITI Aayog – Working Group Estimates 2018

67. https://www.un.org/development/desa/en/news/population/world-population-prospects-2017.html

68. http://hdr.undp.org/en/content/2019-human-development-index-ranking

69. https://tradingeconomics.com/india/employment-in-agriculture-percent-of-total-employment-wb-data.html)

70. http://ficci.in/spdocument/20550/FICCI-agri-Report%2009-03-2015.pdf

71. http://statisticstimes.com/economy/sectorwise-gdp-contribution-of-india.php

72. https://www.ibef.org/industry/agriculture-india.aspx

73. http://agropedia.iitk.ac.in/content/scenario-agriculture-2008-09

74. https://www.domain-b.com/economy/agriculture/20200219_foodgrain.html

75. https://dea.gov.in/sites/default/files/India_External_Debt201314E.pdf

76. https://pib.gov.in/PressReleseDetailm.aspx?PRID=1603539#:~:text=Tota l%20Oilseeds%20production%20in%20the,than%20the%20average%20 oilseeds%20production

77. https://www.ibef.org/news/indias-fish-production-pegged-at-126-million-tonnes-in-1718

78. https://www.domain-b.com/economy/agriculture/20200219_foodgrain.html

79. https://en.wikipedia.org/wiki/Land_use_statistics_by_country

80. whttps://www.researchandmarkets.com/reports/4763057/indian-cold-chain-market-industry-trends share?utm_source=CI&utm_medium=PressRelease&utm_code=s887d6& utm_campaign=1276031+-+Indian+Cold+Chain+Market%3a+Industry+Trends%2c+Share%2c+Siz e%2c+Growth%2c+Opportunity+and+Forecasts%2c+2013-2018+%26+2019-2024&utm_exec=chdo54prd

81. Business Standard. Lucknow, 23-3-2019

82. https://www.indianmirror.com/indian-industries/2019/food-2019.html

83. https://www.goodreturns.in/news/union-budget-2020-benefits-to-india-s-agriculture-sector-1127923.html

84. https://www.rbi.org.in/scripts/AboutUsDisplay.aspx?pg=RegionalRuralB anks.htm

85. https://financialservices.gov.in/list-rrbs-functioning-country

86. https://madhavuniversity.edu.in/rrb-of-india.html

87. https://www.ibef.org/news/indian-agrochemicals-market-to-reach-us-63-billion-by-fy2020-report

88. https://www.ibef.org/news/indian-agrochemicals-market-to-reach-us-63-billion-by-fy2020-report

89. https://eands.dacnet.nic.in/Advance_Estimate/4th_Adv_Estimates2019-20_Eng.pdf

90. https://www.ncbi.nlm.nih.gov/pmc/articles/PMC5268357/#bib0115

91. https://www.thethirdpole.net/2016/08/08/climate-change-will-claim-160000-lives-a-year-in-india-by-2050/

92. https://www.who.int/news-room/fact-sheets/detail/climate-change-and-health

93. https://thewire.in/agriculture/climate-change-agricultural-decline

94. https://eands.dacnet.nic.in/Advance_Estimate/4th_Adv_Estimates2019-20_Eng.pdf

95.   https://thewire.in/agriculture/climate-change-agricultural-decline

96.   https://indianexpress.com/article/explained/reading-the-livestock-census-2019-india-6075270/

97.   https://www.downtoearth.org.in/news/natural-disasters/warning-bell-extreme-weather-is-hitting-india-s-farmlands-harder-72581

98.   https://www.livemint.com/Science/VIlUXxg4fhF9JaYceFsbYI/IMD-to-add-30-Doppler-radars-in-country.html

99.   https://www.downtoearth.org.in/news/more-than-half--farmhouseholds-in-india-are-in-debt-nsso-report-47924

100.   http://ddnews.gov.in/business/exports-agricultural-products-during-march-june-2020-

101.   https://scroll.in/article/903506/why-much-of-india-lacks-access-to-safe-drinking-water-despite-an-ambitious-government-project

102.   https://en.reset.org/blog/water-borne-diseases-india#:~:text=A%20report%20by%20the%20United,of%20water%2Dborne%20diseases%20annually

103.   https://www.icar.org.in/node/119

104.   Wikipedia

105.   Business Standard

106.   India/2018/171http://statisticstimes.com/demographics/country/india-population.php

107.   https://www.cleanindiajournal.com/food-wastage-crisis-in-india/

108.   Financial Express

109.   https://tradingeconomics.com/india/employment-in-agriculture-percent-of-total-employment-wb-data.html

110.   https://www.reportlinker.com/p03607241/?utm_source=PRN

111.   https://www.developmentnews.in/agriculture-rd-spend-reality-check/

112.   https://www.unwomen.org/en/news/stories/2019/7/take-five-rahul-bhatnagar-india